STRUCTURED TECHNIQUES OF SYSTEM ANALYSIS, DESIGN, AND IMPLEMENTATION

STRUCTURED TECHNIQUES OF SYSTEM ANALYSIS, DESIGN, AND IMPLEMENTATION

SITANSU S. MITTRA

TRW Inc.
Boston University

WILEY

A Wiley-Interscience Publication

JOHN WILEY & SONS

New York · Chichester · Brisbane · Toronto · Singapore

Library of Congress Cataloging in Publication Data:

Mittra, Sitansu S.
  Structured techniques of system analysis, design, and implementation / Sitansu S. Mittra.
      p.   cm.
  "A Wiley-Interscience publication."
  Bibliography: p.
  Includes index.
  ISBN 0-471-83081-X : $42.95 (est.)
  1. System analysis.   2. System design.   I. Title.
QA402.M56 1988
003—dc19
ISBN: 0-471-83081-X

87-28577
CIP

Printed in the United States of America

10 9 8 7 6 5 4 3 2 1

To the memory of my parents

Satya Charan Mitra
Champak Lata Mittra

neither of whom lived to see this work

PREFACE

This text provides an in-depth coverage of the structured techniques of system analysis, design, and implementation. The structured method distinguishes between two types of systems—logical and physical. The logical system is a purely conceptual system that provides all the functional capabilities required by the user or customer. The physical system, on the other hand, is the system that implements all these functional capabilities so that the user gets the benefit of them. Accordingly, the structured method insists that the logical system be completely developed before any work on the physical system starts.

The main emphasis of the book is on the structured method and its importance in the development of a system. The philosophy of the structured process is that if the logical system is completely laid out, the physical system becomes quite obvious. Although the layout of the logical system takes a considerable amount of time at the beginning and thereby results in a delay in allowing the user to enjoy the benefits of the proposed system, it saves a considerable amount of time and labor in the long run.

I strongly believe in the structured techniques and use them regularly in my work. I have supervised the design and implementation of several projects, each time supervising a team of six to eight technical staff members. In each case, I used the structured techniques discussed in the book and described in detail in the table of contents. Consequently, I have drawn heavily from my experience in writing this book. In fact, one of the case studies, the on-line financial reporting system, was developed under my direction and supervision and took nearly 15 months. Thus, the book does not describe mere theories, but also illustrates them with real-life examples. A considerable amount of my project management experience has been utilized in preparing this book.

In addition to my industry experience in system analysis, design, and implementation, I have given courses on this subject for the M.S. program in computer science at Boston University and for the B.S. program in computer science at Wentworth Institute of Technology in Boston. In both places, I had to write extensive notes and case study materials in order to teach the subject. I could not find any single book covering the entire subject at an in-depth level. Accordingly, I regard this book as a totally new venture in the field of system analysis, design, and implementation.

The book consists of 15 chapters divided into five parts.

Part I consists of Chapters 1 and 2. Chapter 1 provides the background materials on computer-based information systems and their impact on management. Chapter 2 discusses the system life cycle concept and distinguishes between a logical system and a physical system. The three principal phases of a life cycle—analysis, design, and implementation—are then explored in Parts II, III, and IV.

Part II consists of Chapters 3 through 6. Chapter 3 describes the problem definition and feasibility study for a system, and Chapter 4 illustrates this concept with two case studies: an order processing system and a financial reporting system. This pattern of first introducing the theories and then illustrating them with case studies is carried through Chapter 12. Chapters 5 and 6 introduce the theory and case studies for the system analysis phase.

Part III consists of Chapters 7 through 10. The design phase is divided into two parts: preliminary and detailed. The preliminary system design concentrates on the input and the output of the system along with an overview of the processing. The detailed design elaborates on the latter and concludes with a complete design specification ready to be implemented.

Part IV consists of Chapters 11 and 12. These chapters describe the phase involving the structured programming methodology along with many other related issues, such as preparation of physical site, system conversion, user training, and documentation.

Part V consists of Chapters 13, 14, and 15. Chapters 13 and 14 describe the roles of decision support systems and database systems in the development of computer-based application systems. Chapter 15 addresses a set of heterogeneous topics in system development such as the information center concept, consultant versus in-house expertise in system development, third party leasing, on-line transaction processing, and system development for expert systems.

The overall treatment has been done at a fairly elementary and descriptive level. The reader must be familiar with data processing principles and must have programmed in at least one high-level language such as COBOL, FORTRAN, BASIC, or PASCAL. Some knowledge of college algebra will be helpful in understanding the space estimate techniques discussed in Chapter 9.

The book is intended for *two* types of users:

1. Business or computer science students who want to learn the structured methods of analysis, design, and implementation of information systems

2. Computer system professionals (e.g., systems analysts, information systems specialists) who want to build a system or are involved in the process of building a system

For the first group the book can be used as a text for a one- or two-semester graduate course on system analysis and design. The book contains more materials than can be covered in-depth in a one-semester course. This allows the instructor a considerable amount of choice in selecting topics to teach.

For the second group the book provides instructions on how to build an information system by using structured methods and as such can be used both as a step-by-step procedure manual and as a reference.

Due to the increasing popularity of structured system development techniques among business managers, many colleges and universities are offering courses on these methods. At the same time the number of computer professionals in industry working with structured system development processes is increasing rapidly. Accordingly, the target audience for the book appears to be substantial.

During the time that I was writing the book I received continuous support from my family, my wife Pranati being the cheerleader of the team. My sons, Partha and Ansu, were somewhat appreciative at this time of the continuing pressure of writing a book after having seen me survive the same ordeal while writing my previous book.

I acknowledge the help I received from my former students at Boston University and at Wentworth Institute of Technology in shaping my ideas about structured methodology. I have benefitted from some of the term papers they wrote as part of their course requirements.

It is a pleasure to acknowledge the friendly support of Maria Taylor, and I sincerely thank the staff of John Wiley & Sons for making this production job a success.

SITANSU S. MITTRA

Medfield, Massachusetts
January 1988

ABOUT THE AUTHOR

Sitansu S. Mittra currently works as the Manager of System Planning and Software Maintenance for TRW, Inc. in Lexington, MA. He also holds an appointment as Adjunct Assistant Professor of Computer Science with Boston University.

Dr. Mittra has two master's and a doctorate degree in mathematics from University of Calcutta, India, University of Toronto, and Lehigh University respectively. His areas of interest are: structured system development methodology, database management systems, decision support systems, mathematical modeling, and expert systems. Prior to joining TRW, Inc., he worked as a Senior System Specialist with Unisys Corporation at Cambridge, MA. His previous book, Decision Support Systems: Tools and Techniques, was published by John Wiley. So far he has published over 45 technical papers and reports in computer systems, operations research, mathematics, and artificial intelligence in various professional journals and in-house publications. Currently, he is working on his third book, Principles of Relational Database Systems.

CONTENTS

6 System Analysis: Case Studies   104

PART III
STRUCTURED DESIGN

7 Preliminary System Design: Theory   133

PART IV
STRUCTURED IMPLEMENTATION

11 System Implementation, Maintenance, and Evaluation: Theory 253

STRUCTURED TECHNIQUES OF SYSTEM ANALYSIS, DESIGN, AND IMPLEMENTATION

INTRODUCTION

Part I consists of Chapters 1 and 2. Chapter 1 provides the background materials on computer-based information systems. It discusses the three types of management functions and their respective impact on an organization. MIS in an organization appears as a concept or principle whereby all existing application systems are integrated. Chapter 2 describes the system life cycle concept and distinguishes between a logical system and a physical system. Under structured methodology the logical system is completely developed before the physical system is started.

1

Computer-Based Information Systems

1.1 CONCEPT OF A SYSTEM

In our everyday life we frequently use the word "system" to mean a wide variety of things. On our way to office in the morning we form a part of the *traffic system*. Our children in schools and colleges participate in the *American education system*. In most jobs the fringe benefits require that the employer make a contribution to the health insurance package and the retirement fund of each employee. As a result, the employees can take advantage of the *health care system* and reap the benefits of the *retirement system*. As consumers we use products made by different companies from their *manufacturing systems*. As law abiding citizens we demand that the *criminal justice system* should protect us from crime and should punish the criminals. People holding regular jobs receive their paychecks generated by the *payroll system*. Our monthly utility bills and credit card statements are produced by *billing systems*.

The word "system" is completely meaningful in each of the above examples, and we can think of numerous other examples of usages of system. We are, therefore, forced to ask the question, *What is a system?* The answer is, whatever characteristics are common to all the above situations will define a system. The answer sounds like an oversimplification, but is true.

Alexander ([1], p. 4) has given 10 different definitions of a system, all of which are valid. We accept the following, general definition of a system:

A *system* is a group of elements, either physical or non-physical in nature, that exhibit a set of interrelations among themselves and interact toward one or more goals, objectives, or ends. The elements that comprise a system may be of several different types.

3

The environment is external to the system. It includes everything that is beyond the system's control and thus forms the boundary of the system. Of course, a system and its boundary are interrelated since a system cannot exist in vacuum. The elements of the system are internal to the system. The most significant aspect of a system is the interaction of its constituent elements to achieve one or more preestablished goals. Normally we separate a system into smaller subsystems. The nesting of smaller subsystems within larger ones forms a hierarchy that is characteristic of any system.

A system and all its subsystems accept input and produce output. The goals of a system are directly related to its output. The actual production of the output need not be unique. In other words, given a system, its output can be produced in more than one way. For example, the goal of a traffic system is to ensure safe flow of traffic and avoidance of accidents. The input to this system is the stream of traffic (e.g., cars, pedestrians), and the output is, for example, the same stream moving without accidents at major street intersections. This output can be produced by installation of traffic lights or stop signs at the intersections or by the deployment of traffic police at those places. As long as the ultimate goal of safe flow of traffic is realized, the actual means used are immaterial.

1.2 ROLE OF AN INFORMATION SYSTEM IN AN ORGANIZATION

Our discussion of a general system can be applied to describe an information system. Lucas ([3], p. 4) has defined an information system as follows:

> We define an information system as a set of organized procedures that, when executed, provide information for decision making and/or control of the organization.

Let us compare the above definition of an information system with that of a system given in Section 1.1. We notice that the broad definition of a system has been narrowed down so as to apply to an information system. The following table shows this correspondence (see [4], p. 30):

Definition of System	Definition of Information System
1. Group of elements	Data used in setting up the organized procedures
2. Set of interrelations	Flow of data through the organization and how this flow is related to the operation and control of the organization
3. One or more goals	Well-informed managers equipped with adequate information for decision making and thereby making the organization more efficient

An effective information system captures data as close to its point of origin as possible and then channels them to the information processing stations where they are arranged, calculated, summarized, and otherwise prepared for communication to the decision makers. In most agencies, the information system is at least partially automated. This means that some amount of the data collection, information processing, and generation of reports is done by computers, whereas the rest is handled manually by human beings and mechanical devices such as calculating machines and typewriters.

In an information system, there is a basic objective that justifies the system's existence. For example, the objective of an inventory control system is to ensure that all data related to the inventory of items are captured and processed properly so that a balance is maintained between the two often contradictory goals:

□ To minimize the amount of capital tied in inventory
□ To minimize the occurrences of backordering customer's orders

Different people view the information system differently, somewhat similar to the legendary experiences of six blind men describing an elephant. To a user or customer, the system is a vehicle for generating periodic reports, both detailed and summary, which different levels of management can use in their daily functions. In addition, the system enables the management to access the data files through ad hoc queries in order to gather desired information that is not available through the periodic reports. To a systems analyst, the system represents the logical flow of data through the basic computer cycle of input-process-output, whereby it can collect the requisite data, store and process them to generate reports, and finally disseminate these reports. To a programmer, the system is much more restricted in that it is reduced to individual modules that achieve some very specific functions such as updating a master file and printing invoices using sales receipts. To the computer operations staff, the system consists of detailed operations documentation that specifies how the system is run, how data should be entered and verified, what are the control procedures and restart procedures, and so on.

Any information system thus exists within the confines of an organization, and its users are the employees of that organization. Accordingly, the philosophy and operating principles of the organization have an impact on its information system. The managers of the organization determine their managerial functions and style in accordance with the guiding policies of the agency.

The use of the information system by the managers and the dissipation of the information acquired thereby depend on the style of management within the organization. This brings us to the discussion of the management style practiced and supported in an agency.

1.3 MANAGEMENT FUNCTIONS AND STYLE

In any organization there are three levels of management:

□ Strategic or top level
□ Tactical or middle level
□ Operational or bottom level

Generally, the bottom-level management deals with operational information, whereas the middle and the top levels are concerned with tactical and strategic information, respectively. The required information should be more and more summarized as the level of managers who use them rises higher and higher in the three-tier hierarchy of management. The information system in the organization must keep track of this varying information needs. Figure 1-1 illustrates the three levels of information needs.

The time spans for the reports addressing the needs of the three levels of management differ significantly. The strategic level of managers are concerned with the overall goals and planning for the company and want to compare the company's performance with that of its competitors. So their information needs normally span from 1 to 5 years. The level of uncertainty in these reports remains high. The tactical level of managers are involved with technical information and usually deal with a time span of 1 year. Accordingly, the level of uncertainty in such reports becomes lower. The operational level of managers (e.g., foremen, first line supervisors) implement the work plan on a daily basis and as such deal with information having very little uncertainty. Figure 1-2 summarizes the essential characteristics of the types of information needed by the three levels of management.

Top (strategic) level	Summary reports Goal and planning oriented
Middle (tactical) level	Detailed technical reports Planning, control, and solution oriented
Bottom (operational) level	Detailed operational reports Daily operation and implementation oriented

FIGURE 1-1 Three levels of information needs. (Reproduced with permission from S.S. Mittra, *Decision Support Systems*, Wiley, New York, 1986.)

Management Level	Information Type	Time Span	Level of Uncertainty
Top	Strategic	1–5 years	High
Middle	Tactical	1 year	Medium
Bottom	Operational	Daily	Low

FIGURE 1-2 Information need characteristics. (Reproduced with permission from S.S. Mittra, *Decision Support Systems*, Wiley, New York, 1986.)

The managerial style of the managers depends on two factors: the personal preference of the manager and the organizational guidelines. In a company that practices a democratic and participatory style of management, the managers tend to share information with their subordinates, and the flow of information is both horizontal and vertical. On the other hand, if a company follows a rigid authoritarian policy, the managers tend to be highly close-mouthed and nonsharing. Subordinates cannot get the information available to their superiors. The flow of information remains primarily horizontal and seldom moves vertically downward. In essence, then, a manager's style of management is influenced more by the company policy than by his or her personal preference. An information system's usefulness in dissipating information within an organization is highly dependent on the managerial styles of its managers.

1.4 ORGANIZATION CHART

The organization chart of a company is a graphic representation of its organization structure. As such, it provides a lot of insight into the management style followed in the company. The chart shows the environment in which an information system exists.

In any company there normally exist two types of organizations: formal and informal. The formal organization is what appears on the organization chart, and usually there are standards and procedures that describe the structure of this formal organization. The informal organization is the pattern of relations and groups among members of the formal organization that is not specified on the organization chart. It represents social interactions that exist within the company. The informal organization is a more realistic portrayal of the formal organization since it reflects how people actually interact. For example, it shows which of the several approaches to cutting red tape and bypassing standard procedures is used by employees. Consequently, we should be careful in designing information systems that strictly follow the formal organization chart and ignores the underlying informal organization. The only difficulty is that no formal documentations exist describing the informal organization. It depends on the personalities of specific individuals and patterns of behavior that have developed over time.

The organization chart also shows whether a company has a top-heavy administration, which is totally undesirable. An organization chart should have a pyramidal structure, appearing broad at the bottom and getting progressively narrower as the chart moves upward. Comparing the chart to a tree, we can say that it should look like a tall tree rather than a bushy one. Figures 1-3 and 1-4 show, respectively, a normal and a top-heavy organization chart.

The director of an information system in a company should have the rank equivalent to a vice president and should report directly to the president of the company in a staff responsibility; this provides proper authority to the position. Traditionally, however, this role had been made subordinate to the controller or the vice president of finance, because the data processing functions have primarily been accounting in nature. But with the increased importance of information systems in a company, this trend is changing and the director of information system is becoming more visible in the organization chart.

1.5   MANAGEMENT INFORMATION SYSTEM

The word "system," even in a computer environment, has two broad connotations: It can mean an application system such as payroll, inventory control, general ledger, and so on, or it can mean a supporting software that provides a smooth operation of the computer environment. Examples of the latter are operating systems, database management systems (DBMS), and forms-generating systems. In this book, we shall use the expression "information system," to mean an application system, i.e., a system that is designed for generating a set of reports and screen displays according to a set of procedures and that uses supporting software such as screen formatters and database management systems for more efficient operation.

The *management information system (MIS)* in an organization provides a broad umbrella that includes all the application systems and the decision support system (DSS), if any. However, the MIS does not include the operating system or the database management system, because they are supporting software for running the MIS just as the computer equipment is the supporting hardware for the MIS.

We, therefore, conclude that the operating system and the database management fall outside the realm of MIS, since each is a different kind of system. In an organization, the managers are interested in the capabilities provided by an MIS. If the latter includes a DSS, then the managers get extra support from the DSS in improving their decision-making power. Hence we now explore the nature and contents of MIS.

The MIS in an organization is a collection of all the application subsystems (e.g., inventory, accounts payable, sales forecast, corporate planning) and the decision support system, if one exists. The MIS is thus a philosophy or

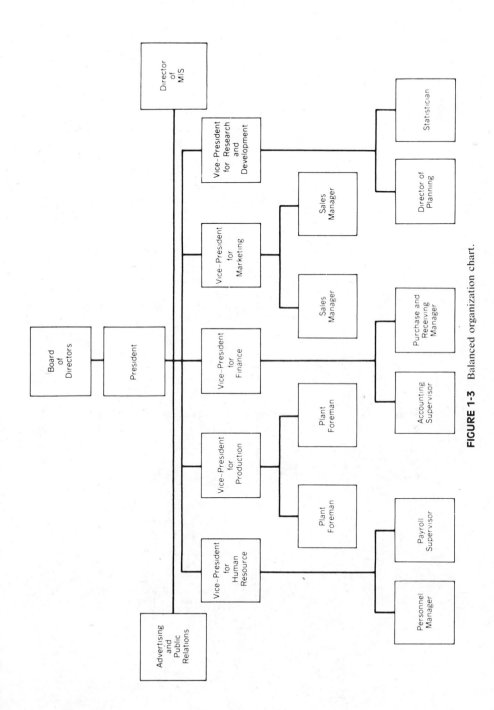

FIGURE 1-3 Balanced organization chart.

9

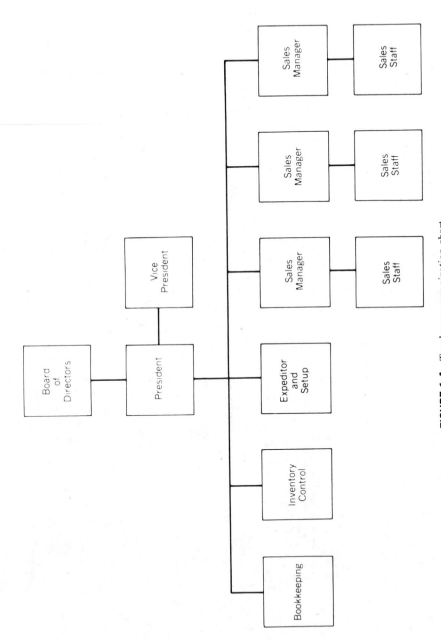

FIGURE 1-4 Top-heavy organization chart.

viewpoint that integrates all the existing systems in the company and makes information available to managers in a readily accessible form. A true MIS should be supported by a versatile DBMS providing a nonprocedural query language capability to retrieve data from the integrated database.

The MIS develops in response to the needs of management for accurate, timely, and meaningful data in order to plan, analyze, and control the organization's activities and thereby maximize its survival and growth. The MIS accomplishes this objective by providing means for input, processing, and output of data, along with a feedback decision-making capability that helps management respond to current and future changes in the internal and external environment of the organization.

A well-developed MIS in an organization should be able to respond to the following questions:

1. What data are needed?
2. When are they needed?
3. Who needs them?
4. Where are they needed?
5. Why are they needed?
6. How much do they cost?

This list is often referred to as 5Ws and 1H.

The MIS combines diverse fields such as computer science, mathematics, statistics, operations research, organization theory, interpersonal skills, and human psychology. It enables people with a variety of skills to work together toward the short-term and long-term goals of the organization. Siegel has described an MIS as follows ([5], p. 31):

An MIS consists of a data bank foundation upon which is built four user-oriented levels—planning, control, forecasting, and modeling—and two technological levels—computing and data administration. Each level is related to a different type of user.

Managers

The *planner* develops strategic long- and short-range plans.

The *supervisor* supports the plan by exercising control; he keeps track of deviations from the plan and brings results into closer conformity to the plans.

Management Specialists

The *forecaster* forecasts the future state of the environment system as well as that of the operation and control systems in order to provide a firm foundation for planning and control.

The *modeler* develops rigorous models based upon the ideas of and for the use of the planner, supervisor, and forecaster.

Information Specialists

The *computerist* analyzes specific applications and writes the programs which serve the needs of the planner, supervisor, forecaster, and modeler.

The *data administrator* defines, creates, and maintains all the data in the information system for the common good of all the users.

Figure 1-5, reproduced from Siegel ([5], p. 30) shows the six levels of MIS.

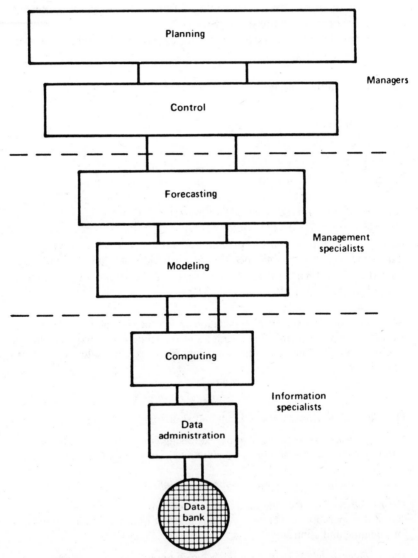

FIGURE 1-5 Six levels of MIS. (Reproduced with permission from Paul Siegel, *Strategic Planning of Management Information System*, Petrocelli Books, Princeton, NJ, 1975.)

1.6 EVOLUTION OF MIS IN AN ORGANIZATION

We have defined the MIS in an organization as a philosophy or principle that takes a global view of the total information needs of a company and integrates the various application subsystems in a unified approach. Accordingly, the development of MIS is an evolutionary process.

Initially, the MIS gets a modest start and addresses a small set of management needs. Gradually, more and more demands are made on the system, and these demands are addressed in a patchwork fashion. Eventually, the system gets too many patches, and then the need arises to integrate all the components in a well-coordinated fashion. Formal, semiformal, and even informal reporting habits are standardized, definite procedures are set up, and schedules and priorities are established. Information flows are structured for early warning of problems, quick response to crises, and clear channels for communicating management directives. Finally, the organization gets its MIS.

Bocchino ([2], p. 7) has written that an MIS

develops in response to the needs of management for accurate, timely, and meaningful data in order to plan, analyze, and control the organization's activities and thereby optimize its survival and growth. The MIS accomplishes this mission for management by providing means for input, processing, and output of data plus a feedback decision network that helps management respond to current and future changes in the internal and external environment of the organization.

Graphically we can represent an MIS as a hierarchy of systems. Let us assume that a company has the following major subdivisions:

a. Finance
b. Production
c. Marketing
d. Personnel
e. Research and Development
f. Decision Support System

Each subdivision uses one or more application systems in its daily activities. For example, finance uses accounts payable and accounts receivable systems, and production uses inventory control and purchasing/receiving systems. Historically, these systems start at different times, as the needs arise and are supported by isolated files with overlapping data elements. Eventually, someone higher up in the company feels the necessity to integrate all these related but isolated systems in order to achieve efficiency and better support for management's information needs. As a result, the MIS is started in the company, whereby the different systems are combined, the isolated

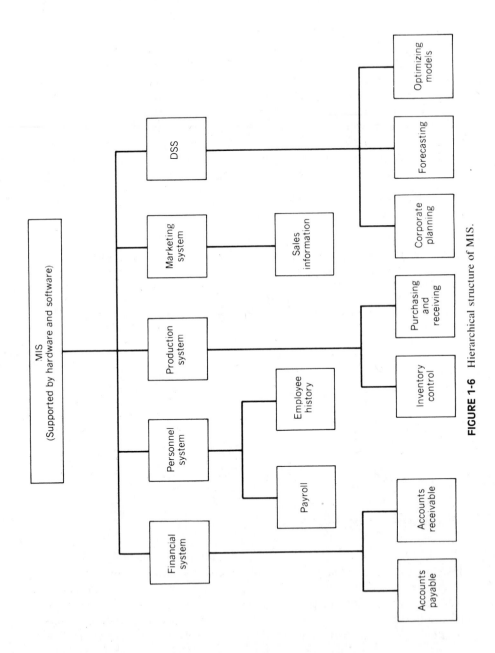

FIGURE 1-6 Hierarchical structure of MIS.

data files are consolidated, and a centralized control is established with the director of MIS being in charge. Figure 1-6 shows the hierarchical structure of this MIS.

1.7 IMPACT OF COMPUTERS

Each application system in an organization forms a part of the hierarchy of systems that is the MIS. As such, computers play a major role in the MIS operation within a company. All levels of employees use the MIS in their daily activities and are thereby affected by the computer. Since the MIS department functions in a service capacity, it establishes relationships with all other departments. The employee reaction to computers is at best a mixed one. Often at the beginning of an automation effort those employees who are going to be affected view the company as an adversary. Not too long ago many of the top-level managers looked upon the computer as a competitor for their individual authority. With the growing trend of computer literacy, this situation is changing. More and more managers are now realizing that computers actually help them in their jobs. Since the computer is just a machine, which is incapable of making decisions despite the rapid progress of artificial intelligence, managers should not feel threatened by them. No computer can ever replace the decision-making authority of a manager. An application system aided by the computer merely provides rapid access to large volumes of data and produces reports at an enormously fast rate.

Computers make a profound impact on people because of the following reasons:

a. Computers can store a virtually unlimited amount of data.

b. Computers do routine repetitive works at extremely high speed. Such works include routine calculations as well as access and retrieval of stored data.

Human beings are far less efficient than computers in both of the above areas.

Historically, when computers were first invented, only *batch applications* were used. Only since the early 1980s has emphasis been placed on interactive applications. An *interactive application* is always characterized by a human/computer interface. It has three parts: input, processing, and output. The processing belongs more to the interactive systems, whereas input and output belong to the interactive applications.

The human factor plays a dominant role in interactive applications. The analysis and design phases of such applications should address this issue. As such, the design of user interface is more an art than a science.

Human factors have not been a traditional concern in the formal study of computers. The main emphasis in the past had been on the optimization of computer time and memory. Program efficiency was the highest goal. Now with less expensive hardware more attention is devoted to user satisfaction rather than computer efficiency.

User–computer dialogue plays a vital role in the design of interactive applications. The principle is that the language of user interaction should be human oriented rather than slanted toward the computer. Accordingly, the following design considerations are important:

- ☐ Give feedback.
- ☐ Help the user learn the system by providing enough prompts and messages.
- ☐ Allow backup and accommodate errors.
- ☐ Control response time.
- ☐ Design for consistency.
- ☐ Structure the display.

1.8 CONVENTIONAL VERSUS STRUCTURED TECHNIQUES OF SYSTEMS ANALYSIS

As a formal discipline, systems analysis is fairly new. Even in the mid-1970s, many good textbooks for the subject did not exist. The whole concept of the analysis and design of systems was at best a vague one. Then there gradually evolved the method that is now regarded as the conventional systems analysis and design process and as such is contrasted with the better and more recent methodology of structured systems analysis and design. However, the difference between the conventional and the structured techniques is one of degree only, not kind.

The *system analysis* process is concerned with analyzing existing systems, determining the information requirements for the organization, evaluating the benefits to be derived through computerization of management systems, and comparing these benefits with the estimated costs for implementing the system. The *system design* process involves the design of new systems and the development of system specifications for programming. The system analysis phase produces the logical design for the new system, whereas the system design phase produces the physical design. Both conventional and structured techniques accept these definitions.

Conventional system analysis starts by analyzing the organization chart and determining the total output requirements, for example, who receives what report and when or how often. It then matches the input data available in the organization with the output requirements and concentrates on the structure of files to collect and store the input data. The system design phase involves essentially the detailed program specifications for implementing the proposed system. The conventional method uses a top-down approach in a limited way in that it examines the requirements of the total system and works out the detail more and more as the design phase progresses. But the logical design and the physical design are not always kept strictly separate.

The emphasis in structured system analysis and design is on the strict separation of the logical design from the physical design. The philosophy of the structured approach is that the conceptual or the logical system must be completely specified and documented during the analysis phase. No reference should be made to physical implementation. Accordingly, the structured system analysis uses the data flow diagram instead of the system flowchart as its graphic tool, because the former is independent of any implementation considerations whereas the latter is implementation bound. The structured system design starts with the logical system prepared as the end product of the system analysis phase and converts it into a physical system. It changes the data flow diagrams into implementation-bound system flowcharts and develops design specifications first at system level and then at program level. Thus, the development of the complete logical system is a prerequisite for the development of the corresponding physical system. There is a total separation of the logical system from the physical system.

The structured method is rapidly gaining popularity in industry because of its superiority over the conventional process. In the structured approach, considerably more time is devoted to the development of the conceptual system and no programming is done until the physical system design phase. As a result, the programs are fairly straightforward and error-free. Also, documentation is easier to generate. The extra time and cost incurred due to the initial analysis and design phases are amply made up by the relative ease and speed of the programming process.

1.9 SUMMARY

The chapter began by defining what a general system is and showed how that definition is modified to apply to an information system. Different types of computer-based system (e.g., operating system, DBMS, application systems) were described. The role of managers as the end users of a management information system was emphasized. It was pointed out repeatedly that a computer information system cannot replace a manager nor can it strip a manager of the human decision-making authority.

Next, the chapter discussed the evolutionary nature of the MIS in an organization and how it can benefit the managers at all three levels—top, middle, and bottom. The view of MIS as a concept or a principle whereby all the existing application systems are integrated under the board umbrella of MIS becomes predominant.

Since computers are the vehicles for implementing an information system, they play a profound role in an organization. With the advancement of technology, the hardware has become less expensive and interactive applications are becoming more popular than batch applications. As a result, human factors are becoming more importance in the design of an information system. User–computer dialogue now plays a vital role in interactive systems.

The chapter closed with a discussion of the conventional versus structured methods of system analysis and design. Both methods are based on the following principles:

a. The logical design must be complete before the physical design.

b. The logical design takes place during the analysis phase, whereas the physical design is done during the design phase.

However, the conventional method does not keep the logical design strictly separate from the physical design, whereas the structured method maintains the separation very rigorously. The structured method is gaining popularity due to its superiority over the conventional method.

1.10   KEY WORDS

The following key words are used in this chapter:

batch application

conventional system analysis and design

hierarchy of systems

information system

interactive applications

management information system (MIS)

operational level of management

organization

organization chart

strategic level of management

structured system analysis and design

system

tactical level of management

user–computer dialogue

REFERENCES

1. M. J. Alexander, *Information System Analysis*, SRA, Chicago, 1974.

2. William A. Bocchino, *Management Information System: Tools and Techniques*, Prentice-Hall, Englewood Cliffs, NJ, 1972.

3. Henry C. Lucas, Jr., *The Analysis, Design, and Implementation of Information Systems*, McGraw-Hill, New York, 1985.

4. Sitansu S. Mittra, "Information System Analysis and Design," *Journal of Systems Management*, April 1983, pp. 30–34.

5. Paul Siegel, *Strategic Planning of Management Information System*, Petrocelli Books, Princeton, NJ, 1975.

REVIEW QUESTIONS

1. Define a system. Take any social system (e.g., health care delivery system) and identify the elements comprising the system, their interrelationship, and the common goals of the system.

2. Describe how an operating system is different from a payroll system. Are there any similarities between them?

3. Explain the evolutionary process involved in the development of an MIS in an organization.

4. Take an organization, say the company where you work. Identify the kinds of problems that are handled by the three levels of management, and indicate their respective time spans. Your answer should be specific to the company, not general in nature.

5. Define an organization chart. Distinguish between a regular and a top-heavy chart.

6. What is meant by the list of 5Ws and 1H?

7. Explain the concept of MIS. What are the six levels of MIS?

8. Distinguish between a batch application and an interactive application. Why does the user–computer dialogue play a vital role in the latter?

9. Distinguish between the conventional method and the structured method of system analysis and design. Which one is superior and why?

10. What are meant by the logical design and the physical design?

11. Are top-down design and structured design synonymous? (The answer is *not* in the book.)

12. Does programming belong to the logical design phase or to the physical design phase or to neither? (The explicit answer is *not* given in the book.)

2

System Life Cycle

2.1 STRUCTURED METHOD OF SYSTEM STUDY

In Section 1.8 we discussed the two principal approaches to system study: the conventional method and the structured method. The rest of this book will be devoted to the structured method of system analysis, design, and implementation.

The structured method is sometimes confused with the top-down method of analysis and design. Under the top-down method, a global view of the total system is taken; that is, the system team looks at the big picture. Then the system development proceeds from the more general top-level requirements to the more detailed lower-level requirements. During this process, however, the analysis phase (i.e., *what* functions are provided by the system) is often mixed with the design phase (i.e., *how* the requirements are met). As a result, backtracking may prove to be difficult.

The structured method follows the same top-down approach in that it looks at the overall system requirements. However, it keeps the analysis and design phases strictly separate. Consequently, the backtracking becomes comparatively simple. If the customer wants to add a new requirement or modify an existing requirement during the analysis phase, the incorporation of that change into the logical system becomes fairly straightforward. At any rate, it does not involve the more costly process of changing one or more programs that have already been implemented. This method requires that during the analysis phase the logical system is specified; that is, a complete specification of the functional capabilities of the system is provided. During the design phase the detailed physical system is specified on paper. The implementation phase then converts this paper system into a computer-based system.

The structured method is decidedly superior to the conventional method because it completely specifies the logical system before going into the physical system. The customer is thereby made completely aware of the exact functional capabilities of the proposed system before any design work has actually taken place.

2.2 PRINCIPAL PHASES OF SYSTEM DEVELOPMENT

There are five principal phases of system development:

a. Problem definition and feasibility study
b. System analysis
c. Preliminary system design
d. Detailed system design
e. System implementation, maintenance, and evaluation

These five phases are grouped into three logical categories: system analysis consisting of (a) and (b), system design consisting of (c) and (d), and system implementation consisting of (e). Clearly, the distinction between phases (a) and (b) is one of degree, not of kind. In other words, phase (a) starts at a superficial level that is further pursued and detailed in phase (b). A similar comment applies to phases (c) and (d). The only distinction between phases (d) and (e) is that in (d) the physical system is designed on paper, whereas in (e) that "paper" system is converted into electronics.

The entire system development process is interactive in nature. As each phase concludes, the system team prepares a report containing findings and recommendations as the end product of that phase. The customer reviews the report and provides suggestions and/or corrections. Once the report is approved by the customer, work starts on the next phase. Thus, a continuous dialogue must be ongoing between the customer and the system team during the entire system development process. The following table summarizes the main items that should be included in the end products of the five phases:

Phase	End Product
1. Problem definition and feasibility study	★ Scope and objectives of the study ★ Description of proposed system ★ High-level data flow diagram ★ Feasibility issues
2. System analysis	★ Deficiencies of existing system ★ Functional capabilities of the proposed system

Phase	End Product
	★ Detailed data flow diagrams
	★ Data dictionary for processes, data flows, and data stores
	★ Cost/benefit analysis
3. Preliminary system design	★ Flowcharts of proposed system
	★ Input and output of proposed system
	★ Screen formats
	★ Alternative solutions and recommendations
4. Detailed system design	★ Record and file structure
	★ Auxiliary storage estimate
	★ Schema design (if applicable)
	★ Data communication network and traffic volume (if applicable)
	★ Structure charts
	★ Program flowcharts or input processing output chart
	★ Equipment specification
	★ Personnel selection
	★ Detailed cost estimate
	★ Implementation plan
5. System implementation, maintenance, and evaluation	★ Structured coding
	★ Testing plan for programs
	★ User training
	★ System documentation manuals
	★ System backup, recovery, and audit trail procedures
	★ Maintenance procedures for computer operations staff
	★ Evaluation plan

The final goal of the system development process is an operational application system. Such a system normally appears as a terminal node in the hierarchical tree structure of an MIS (see Figure 1-6). Alternatively, system development technique can be used for system integration. This involves the process of effectively interfacing computer hardware, software, and data communication subsystems to achieve a cohesive information processing resource that is responsive to the needs of the organization. In the early 1960s, the system integration was almost entirely hardware oriented. However, with the rapid technological changes of the 1970s, the system integration has emerged as a distinct discipline that uses the five-phase system development method. The system integrator now assumes the role of system developer and provides the best available cost-effective solution regardless of the computer vendor.

2.3 CYCLIC NATURE OF SYSTEM DEVELOPMENT PROCESS

Structured system analysis and design are directly related to the system life cycle. The latter consists of the five phases mentioned in Section 2.2. Each phase results in an end product (i.e., a report), which leads to the next phase. The first phase is problem definition and feasibility in which a system is developed in order to address a problem. For example, a company may find that due to an expansion of its workforce, the manual payroll system is causing too many errors and thereby results in employee resentment. A manufacturing firm may find that due to an obsolete inventory control system too many customers are backordered on their requests. This may lead to a bad public image and an eventual loss of customers. A subscription agency may get too many calls or letters from its irate customers, who claim they are being billed for amounts already paid. In fact, they may even produce canceled checks to show that the payments were indeed made. All these situations indicate the existence of problems. Such problems can be addressed by implementing a payroll system, or an updated inventory system, or an invoice system. Thus, problem definition indeed initiates the system life cycle.

As the system development proceeds through the successive phases of system analysis, preliminary design, detailed design, and implementation, the system life cycle nears its end phase. Then the maintenance of the operational system continues. However, the average life span of a system is 3 to 5 years. Usually, after 3 years or so, some overhauling is required of the system. This may arise due to new requirements from the management, or an upgrading necessitated by some new technology in the form of new hardware or software, or both. Consequently, a new request comes up in the form of a new problem definition. This brings us back to the beginning of the system life cycle, namely, the problem definition phase.

It may seem that the development of a system proceeds smoothly from one phase to the next. Problem definition is followed by system analysis; system analysis is followed by system design; in general, the end of one phase marks the beginning of the next phase. However, in practice, it seldom happens that way. It is quite possible that an error in problem definition is discovered later during the system analysis phase, or a new request by the customer leads to a redefinition of the problem, or due to unforeseen circumstances, the system conversion is delayed during the implementation phase. The important thing is that a sense of progression be maintained from one phase to the next. Figure 2-1 shows the cyclic nature of the system development process.

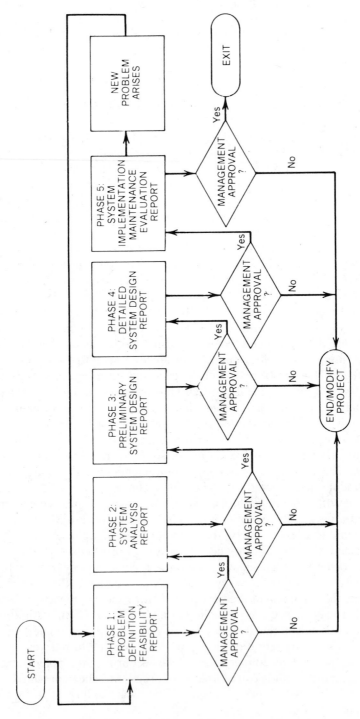

FIGURE 2-1 Cyclic nature of system development.

2.4 THREE LAWS OF SYSTEM ANALYSIS

The structured method of system analysis and design requires that the user be involved at all stages of the system development process. As mentioned in Section 1.8, today's systems often involve interactive components that need user-friendly interfaces. These applications depend on the interaction between the user and the computer. As a result, the system analyst has to play a liaison role between the users and the system technicians. This role does not need any new methodologies. Instead the system analyst must know how to use the existing tools and techniques of system development.

In a recent article, Nakisher [2] has proposed the following three laws of system analysis in order to emphasize the need for user involvement in the system development process:

First Law: You can't listen with your mouth open.
   Corollary 1: Always ask questions.
   Corollary 2: Always listen to the answers.

The implication of these three statements is that the analyst must elicit system requirements from the users instead of telling them what the requirements should be. Nakisher recounts from his experience that once while he was conducting an interview with the representatives of a company's user liaison committee, his colleague, who was an efficient COBOL programmer, started telling the customer what the proposed system should be instead of allowing the customer to explain his requirements. Listening to answers is often more difficult for analysts than asking questions. More often than not, analysts ask a question only to answer it themselves. Thereby, they do not allow the users to express their viewpoints. As a result, the new system may not include all of the user requirements.

Second Law: Never presume the design until the analysis has been completed.

The second law forms the very basis of the structured method, that is, the conceptual or logical system must be designed and approved by the user before the physical system is even attempted. Obviously, an analyst who does not follow the first law cannot follow the second law. To illustrate this point, Nakisher gives an example from his consulting experience where an analyst asked a user if a report should be structured one way or another. When the user rejected both alternatives and wanted the report in another way, the analyst replied, "But that's not the way the system works!" Clearly, the analyst was violating the second law in that he was presuming the physical system before thinking out the logical system.

Third Law: Don't assume.

When analysts assume that they know the answers, they have an illusion of power and thereby cannot learn. With the current changes in user attitudes, analysts not familiar with the structured method can easily get upset with users who are too demanding. On the other hand, when users compromise on user friendliness of a proposed system, the system produced does not meet the user needs; such a system is doomed to failure.

These three laws bring us to the issue of user involvement in the development of a system.

2.5   USER INVOLVEMENT IN SYSTEM DEVELOPMENT

The philosophy behind the structured method of system analysis and design is that a system succeeds when there is heavy user involvement during all the phases. In order to clarify this issue, let us consider two possible scenarios:

Scenario 1

A small manufacturing firm called Product, Inc., has recently been getting lots of complaints from its customers regarding its faulty order processing system. Some customers are needlessly backordered although the firm has the items in its inventory; others are often denied discounts despite their payments within the discount period; still others complain that they have received items that were not ordered. In order to address this problem, the president of the firm has asked a friend of him who has been running a system consulting practice for the past 5 years to design and implement a sound order processing system. Fortified with the backing of the company president, the consultant took a software package that he used earlier for another customer, modified it slightly to suit his friend's needs, and implemented it in Product, Inc. During the design and implementation phases, the consultant did not share his findings and recommendations with the user of the system. After the system implementation, the president made an announcement that the new order processing system will be used from the next week. The employees showed great dissatisfaction with the system.

Scenario 2

Another small manufacturing firm, Electronics Partners, has been plagued with a similar problem as Product, Inc.'s. Its president, being a systems person, requisitioned the services of a data processing service bureau and insisted that his employees be consulted and involved in all the phases of the proposed system development. As a result, Electronics Partners had to wait longer than Product, Inc., to get its order processing system. However, after the system implementation was complete, there was a wide acceptance of the system among the employees.

The above scenarios are hypothetical, but they are useful to emphasize the following points:

a. User involvement in the design and implementation of an information system results in favorable user attitudes. The users regard the system as "our" system as apposed to "their" system or one designed without user involvement.

b. A system designed by an outside team without consulting the users faces tough luck. Instead of regarding the system as a solution to the problems, the users regard it as an imposition and try to avoid it as much as possible. Despite top management directive, the system does not succeed.

Favorable user attitudes toward a system lead to high levels of system use, especially if a system is of high quality. High technical quality should lead to more favorable attitudes and perceptions since a high-quality system provides a better user interface and better output than a low-quality system. High levels of use of an information system make it more likely that a user will take action based on the information provided. Depending on the nature of the analysis, the problem, and the information, high levels of use may lead to high or low levels of performance. In fact, one major criterion to determine if an information system is effective is to measure the volume of voluntary use of the system. If a large number of users use the system on a voluntary basis, that is, without any official mandate from top management, then it is reasonable to conclude that the system is efficient.

There is some disagreement on the ideal amount of user control and user involvement in system analysis and design. Some authors (e.g., [1], p. 116) claim that users should be placed in charge of the system design since they know their present operations and future needs and are in the best position to identify the shortcomings of their current system. Lucas ([1], p. 114) has given two examples where he followed this principle. In one case, a labor union of farm workers actively participated in the system design process and was guided by a team of system consultants consisting of a group of college faculty members and students. In the second case, a group of students in a course helped a small investment company that buys and raises cattle as a tax shelter for its stock holders design a computer-based information system for managing production, accounting, and breeding considerations.

The principle of letting users design their own system can be dangerous. In a small organization the principle may succeed, as it did in the examples cited by Lucas. However, in a large organization with a complex array of activities, this theory does not work. Although user involvement in system analysis and design is essential, we should not stretch it too far. We should allow the user to provide input during the system analysis phase, i.e., when the logical system is being produced, to critique the proposed system if it does not satisfy all the user requirements, to help the system designers by identifying the deficiencies of the existing system, and to participate in all planned change activities that constitute the system development process. However, users, in general, are not knowledgeable on the intricacies of information systems theory. Consequently, the practice of putting the users in charge of designing the information system may backfire.

2.6 ROLE OF SYSTEM ANALYST

As has been mentioned, the structured method insists that users be involved heavily in the system analysis and design. This means that someone on the system development team must play the role of a liaison between the users and the system technicians. It is the system analyst who plays this crucial role, as illustrated in Figure 2-2.

There are two levels of communication that a system analyst must perform:

a. Communication with the users
b. Communication with the system technicians

The communication with the user is labeled nonsystem oriented because the users are not assumed to be knowledgeable in the technical details of system. Through repeated meetings and interviews, the system analyst learns the detailed requirements of the users. These requirements are almost always expressed in terms of the end products, such as reports, graphics, and databases, that the users want. The system analyst must be able to convert these requirements into system-oriented specifications on a global level.

Once these conversions are complete, the second level of communication starts. This involves a more system-oriented technical dialogue between the system analyst and the system technicians, who include programmers, system designers, and documentation writers. Together with these colleagues the system analyst reviews the technical specifications of the user requirements in order to scope out the logical system. It is also the primary responsibility of system analysts to clarify any misconceptions or vagueness of user requirements, which means they continue to play the liaison role.

FIGURE 2-2 Role of system analyst.

In order to succeed, system analysts must be excellent communicators and technically competent. Without super communication skills, they will be unable to talk with users and get their requirements. In addition, they must translate the requirements into technical specifications for the system technicians. For these reasons, system analysts have to be knowledgeable in a wide variety of subjects such as computer science, mathematics, operations research, organization theory, and human psychology.

When users describe their requirements, the analyst has to determine whether they are reasonable and feasible, at least on a global basis. Even during the analysis or preliminary design phase, the users may request some last minute modifications in their requirements. The system analyst has to decide the consequences of these changes and then assess the ripple effects that the changes will have on the overall system. Unless the analyst is technically competent, he or she will be unable to handle these situations. Of course, on occasions the analyst can say that the user's questions are outside the area of his/her expertise. But that should be an exception rather than a rule. In fact, if the analyst exhibits hesitation or ignorance in response to user requirements most of the time, then the users will soon lose respect.

At the other end of this liaison role are the system technicians. The analyst must convert the user requirements described in plain English into system specifications written in technical computer jargon. This requires technical competence in computer science and possibly other allied fields such as mathematics, and operations research. If a system technician on the team raises objections to the analyst's specifications, the analyst must be able to respond in a technical manner. If the objections are valid, the analyst must concede and realize the mistake; otherwise, the analyst should point out the flaws in the objections and refute them. In either case, technical competence is essential.

2.7 SUMMARY

The chapter began with a reiteration of the superiority of the structured method of system development. Under this method, the analysis and design phases are kept strictly separate. The logical system is completely spelled out before the physical system is started. A total system development process has five phases:

a. Problem definition and feasibility study
b. System analysis
c. Preliminary system design
d. Detailed system design
e. System implementation, maintenance, and evaluation

Each phase has an end product in the form of a report and recommendations for the next phase. The first two phases result in the specification of the logical system. The next two phases produce the physical system on paper. The last phase converts the "paper" system into an electronics system.

The entire system development process is cyclic in nature. It starts with a specific problem and addresses that problem. However, the average life of a system ranges from 3 to 5 years. Consequently, within 3 years or so after implementation, some new problems crop up and the system development starts all over again. Hence the name system life cycle.

The success of the structured method depends to a great extent on user involvement during the analysis and design phases. Nakisher proposed three laws of system analysis that collectively reinforce this concept. System analysts must listen to users to understand their requirement, must keep an open mind, and must never presume the design until the analysis has been completed. They should not impose their own preconceived notions on users.

Continuous user involvement during system development always results in a better and more acceptable system. The system users can identify themselves with the system and do not think of it as being imposed on them by management. This results in a higher degree of use of the system. In fact, one major criterion to determine if an information system is effective is to measure the volume of voluntary use of the system.

Opinion is somewhat divided on the issue of the amount and level of user involvement during the system development. Some authors insist that users should be placed in charge of the entire system design and should guide the system technicians. Since the users know their organization and the shortcomings of the current system, they are the most equipped to guide the system development team. Others believe that users should only play an advisory role in system development, while the real technical work should be undertaken by the system developers.

System analysts play a crucial role in the complete system life cycle. They participate in two-level communication. At one end, they determine user requirements by interfacing with users. At the other end, they communicate with the system technicians and convert the user requirements into system specifications. This dual role requires that system analysts be technically competent as well as excellent communicators at both oral and written transactions.

2.8 KEY WORDS

The following key words are used in this chapter:

analysis phase	structured method
conventional method	system life cycle
design phase	system technicians

feasibility study	three laws of system analysis
logical system	top-down method
physical system	user involvement
problem definition	user requirements
role of system analyst	

REFERENCES

1. Henry C. Lucas, Jr., *The Analysis, Design, and Implementation of Information System,* McGraw Hill, New York, 1985.
2. Warren S. Nakisher, "What's Wrong in System Analysis," *MIS Week,* January 11, 1984.

REVIEW QUESTIONS

1. Explain the concept of system life cycle. What are its five phases?
2. What is meant by the statement, Analysis deals with the *what* issues of a system, whereas design addresses the *how* issues?
3. Both design and implementation phases of a systems study deal with the *how* issues. In what respect then do they differ? Explain the need for each of them.
4. Enumerate the five phases of a systems development process and identify the end products of each phase.
5. What is the distinction between the conventional method and the structured method?
6. Enumerate the three laws of system analysis. Describe an example from your own work experience in which these laws can be applied successfully.
7. Why should users be involved in system development? What are the advantages and disadvantages of user involvement?
8. Explain the dual role played by a system analyst in the system development process.

STRUCTURED ANALYSIS

Part II consists of Chapters 3 through 6. Chapter 3 discusses the theory of problem definition and feasibility study, the first phase of the five-phase system life cycle. It describes the fact gathering process and the three feasibility issues: technical, economic, and operational feasibilities. Chapter 4 illustrates the theory of Chapter 3 through two case studies: an order processing system and a financial reporting system. Chapter 5 addresses the second phase of the system life cycle, system analysis. System analysis examines in more detail the functional capabilities of the proposed system and uses data flow diagrams and a data dictionary to document the findings. Deficiencies of the current system are also determined and documented. Finally, alternative solutions are proposed using the technique of automation boundary. The cost/benefit analysis addresses the vital question: Can the benefits of the proposed system outweigh its cost? Chapter 6 continues the two case studies of Chapter 4 through the system analysis phase.

3

Problem Definition and Feasibility: Theory

3.1 PROBLEM DEFINITION AS STARTING POINT

The origin of any systems study can always be traced to the existence of a problem. Often the user encounters difficulties and asks for help. At other times, a problem may have existed for quite some time and then is identified by management as an area of poor performance in the organization. As a result, a systems study is undertaken. In any case, however, a systems team is assigned the task of looking into the problem area and coming up with a proposed plan of action.

The process of problem definition is iterative in nature. The system analyst, as the liaison between the systems team and the customer, starts with a rough definition of the problem and lets the customer review it and comment upon it. The initial definition of the problem is based on the system analyst's meetings with the customer. In Section 2.4 we discussed the three laws of system analysis. The first law along with its two corollaries form the basis of the problem defining stage; the system analyst listens to the customer's complaints, asks pertinent questions, and gradually perceives the problem. The customer may often describe only the symptoms rather than formulate the underlying problem accurately. For example, a symptom may be continuous loss of old customers of the company and a consequent drop in profit. However, the underlying problem may be a poorly maintained inventory, a faulty customer invoicing system, a poor sales strategy, or a combination of these. The system analyst plays the role of a diagnosing physician in analyzing all the symptoms reported by the client and defining the problem(s) underlying these symptoms.

Based on interviews with the client, the system analyst writes a brief description of his or her understanding of the problem and reviews it with the client. At the initial stage, the users as well as the management of the client company should be involved in the discussion. The users should respond to written statements, ask for clarifications, and then rectify errors and misunderstandings. The end product is a clear statement of the problem. This phase may be brief, lasting a day or two, or may last for weeks. It all depends on the scope and extent of the problem. A misunderstood or ill-defined problem almost guarantees that the system will fail in the future.

3.2 TECHNIQUES TO DEFINE THE PROBLEM

In order to define the problem, the system analyst must analyze and understand the client's organization. An organization chart of the client's company helps in this endeavor. In addition, the analyst has to study the management style and the managerial functions in the company, because the proposed system will be operational in that environment. Usually, it is easier to get user involvement if the managerial style is open and participatory; in a highly structured authoritarian firm such involvement is rarely available.

Two techniques are widely used to gather facts that are necessary for the problem definition:

a. Interview different levels of personnel.
b. Review current operations manual and any available reports or studies done earlier in order to address the problem.

By conducting interviews with different levels of personnel, the analyst gets different perspectives of the existing problem. He or she can also ask for the interviewee's idea about possible solutions. Although each separate interview is fairly subjective in nature, the analyst can form an objective opinion by synthesizing the various viewpoints, sometimes conflicting, of different employees in the company. The analyst can and should interview the same person more than once if there are significant discrepancies.

The current operation manuals describe the operations performed in the company. However, there are occasions when the actual procedure is different from the procedures described in the manuals. In such a case, the analyst must clarify the situation and resolve discrepancies. Interviewing the appropriate personnel is the only way to resolve such differences.

If the problem being addressed had been existing in the company for a long time, there is a possibility that reports of earlier studies exist in the company. By studying these reports, the analyst can get a better understanding of the problem. He or she can also learn from these reports why certain earlier attempts to solve the problem had failed and become better able to avoid certain pitfalls.

The interviewing techniques and the review of current manuals and reports of earlier studies are complementary to each other. Jointly they provide a complete picture. We shall now examine each of these two techniques.

3.3 INTERVIEWING TECHNIQUES

An efficient system analyst must be proficient in interviewing. An interview is a valuable source of information during the problem definition phase. It is essentially an iterative process. The analyst should start by interviewing a small number of supervisory personnel to get the big picture. At the next stage, the analyst wants to fill in the details to get a clear understanding of the problem. For this purpose, he or she interviews the next lower level of personnel, i.e., those in charge of the daily operations. Obviously, the analyst should ask permission from the supervisor to interview the staff. For example, if the tentative problem is related to a payroll operation, the analyst should start by interviewing the vice president of finance, the director of accounting, and the payroll supervisor in order to see how the payroll system impacts the entire organization. At the next round, the analyst should ask the payroll supervisor's permission to interview the payroll clerks in order to understand the day-to-day operation and maintenance of payroll records.

A good analyst can often form a fairly complete picture of the existing or proposed system from the first round of interview with the supervisory personnel. However, questions and ambiguities will invariably arise, and many key pieces of information will be missing. This will necessitate interviewing people directly involved in the system. They can answer questions, clear up the ambiguities, and supply the missing pieces.

There are three stages of an interview process:

a. Interview preparation
b. Interview session
c. Interview follow-up

3.3.1 Interview Preparation

The interview preparation stage precedes the actual interview session. The analyst must structure it as follows:

a. Purpose. The analyst examines the materials already gathered about the organization as a whole and the particular problem. He or she lists a set of objectives for the interview, which includes the clarification of missing pieces and ambiguities.

b. Interviewees. The analyst studies the organization chart in order to identify the persons to be interviewed. As noted earlier, he or she should interview the supervisory personnel at the beginning and then talk to the next level of people. The objectives of an interview are highly dependent on the interviewee; however, there should be some overall objectives that are relevant for each interviewee. In addition, the analyst must know the interviewee's position in the organization chart, as well as basic functions and responsibility.

c. List of questions. Based on the objectives of the interview, the analyst must prepare a list of questions to be asked or discussed during the interview. This list should be specific to the interviewee, although some of the questions would be relevant to several interviewee. The analyst should not regard the list as absolute. If the interview begins straying from the key point, the analyst should be able to consider alternative follow-up questions. Since he/she cannot anticipate everything, the list of questions should be made flexible.

d. Time and place. The analyst should contact the interviewee by phone or by letter and schedule a mutually convenient time and place to conduct the interview. Normally, the interview takes place in the interviewee's office. For future records, it is advisable to put in a short memo addressed to the interviewee with the following items:

☐ Purpose of the interview
☐ General guidelines about the questions to be discussed during the interview
☐ Time and place of the interview

Typically, the analyst calls the interviewee, sets up an appointment, and confirms it in a follow-up memorandum.

3.3.2 Interview Session

A well-conducted interview consists of three separate parts.

a. Opening. The key objective of the opening is to establish rapport. The analyst identifies him- or herself and describes the purpose of the interview. The analyst should share with the interviewee the statement of scope and objectives of the overall project. If necessary, he or she should also indicate the person (e.g., president of the company) who authorized the interview. In fact, the authority of the analyst comes directly from the authority of the person who requested the system study to be done.

In order to relax the atmosphere, some pleasantries may be exchanged at the beginning. However, they should not get out of hand and should not take more than the first 5–10 minutes. After that the real interview process should start.

Analysts should *not* apologize for taking the interviewee's time. They must not think the interviewee is doing them a special favor by consenting to the interview. In fact, the favor is mutual in that the analyst is helping the interviewee by formulating the problem and proposing a solution to the problem.

b. Body of the interview. The body is most important because here the analyst is able to get some specific answers. Since the analyst requested the interview, he or she should start the session. Ideally, the analyst should have a list of specific questions, leading the session by asking a fairly broad question such as, I had some difficulty in understanding this procedure (specify the procedure), can you help me on that? This gets the session started on the right track. The stream of questions and answers normally brings out more and more information. However, occasionally an interview may be futile. In such a situation, the analyst should end the interview quickly and professionally so as to avoid wasting more time.

During the interview the analyst should take notes, especially the key points. However, it should not be overdone. It is not necessary to record every word or to write complete sentences. Abbreviations and key expressions that will later trigger some chain of thought are sufficient. If necessary, the analyst can tape the entire interview. But he or she must get the interviewee's permission beforehand to do so. Many people feel very uncomfortable in getting their comments recorded. In no case should the analyst carry a tape recorder to an interview without getting the prior approval from the interviewee. In any case, the analyst should transcribe a complete proceeding of the interview after it is over. In fact, a good way to end an interview is to make a mental summary of the key issues and articulate them to the interviewee. This gives both parties a chance to make corrections.

The analyst's attitude toward the interview makes it a success or a failure. He or she should listen rather than talk (remember the three laws of system analysis!) and should never try to belittle the interviewee or appear confrontational. The analyst should ask probing questions but should not cross-examine.

c. Closing. At the end of the interview the analyst should thank the interviewee for his or her time and interest and should enumerate the key issues discussed during the session. The analyst should also promise a written follow-up, which will make the transactions of the interview available for review by the interviewee. If the interview runs longer than its scheduled time, the analyst should ask permission to continue or offer to reschedule. If the analyst feels the necessity of coming back for a second interview at a later date, he or she should say so. Usually, an interview is an iterative process since it is seldom that the analyst can get all the information at the first round.

3.3.3 Interview Follow-up

As soon as possible after the interview has ended, the analyst should consult the notes, listen to the tape recording (if there is one), and jog his or her memory in order to prepare a complete transaction of the interview. Then the analyst should prepare a memo addressed to the interviewee, mention the time, place, and topic of the interview, and include the detailed transactions. The analyst should ask for the interviewee's comments and corrections. This practice provides good public relations and establishes adequate documentations for later use.

If a follow-up is necessary and if it takes more than a few minutes, the analyst should ask for an appointment in the memo. Otherwise he or she should carry on the follow-up by telephone and should document the telephone conversations in a memo.

3.4 REVIEW OF MANUALS AND EARLIER STUDIES

The current operating procedures in a company are normally described in a series of appropriate manuals; for example, the method of adding an item to the inventory or deleting it from inventory should be available from the inventory update manuals. If the system study is to address the problem related to inventory management, the analyst must review this manual, as well as related manuals such as purchasing and receiving procedures. All of these manuals provide excellent sources of information for understanding the current procedures and locating the possible problems. It is true that sometimes the actual practices followed in the company somewhat deviate from the procedures described in the manuals. Should this be the case, the analyst should ask for a clarification during the interview with the appropriate personnel.

Another source of information is any report of earlier studies on the same subject. If the problem has been persisting for some time, it is possible that there have been some earlier attempts, perhaps in the form of an in-house study. Even if such studies failed or proved inconclusive, they can provide valuable information and some insight into the problem definition. If there was an in-house study and if the authors of the study are still working in the company, the analyst should schedule interviews with them. Of course, during the interview the analyst should emphasize that the purpose is not to discredit the earlier study but to learn from its findings. This helps the analyst by not alienating the authors of the earlier study.

3.5 OVERVIEW OF PROPOSED LOGICAL SYSTEM

System analysts gather facts by interviewing people and reviewing current operations manuals and earlier studies. They then synthesize all the information and, with the help of other members of the system team, formulate a problem definition on a global basis. This is indeed the starting point of the logical system development. From now on the work is done by the team members on a group basis.

The problem definition states the specific problem at hand and describes the scope of the proposed system. It lists the functional capabilities of the new system. These constitute the preliminary requirements analysis where the team converts the user requirements into more technical specifications. The problem definition should include the following items:

a. **Problem statement.** Why does the user need a new system or want to modify the current system?

b. **Objectives.** What will the new system achieve; i.e., what are the functional capabilities of the new system?

c. **System boundaries.** What should be included in the new system and where will it interface with other systems?

d. **Preliminary ideas.** What is one possible solution of the existing problem?

e. **Project cost.** How much will it cost to implement the new system?

The last item, project cost, is only an estimate and far from being accurate. In fact, it is quite acceptable at this preliminary stage to say that the estimated costs can vary within a range of ± 50 percent. These figures merely give the user some gross estimates of the cost.

3.6 SUBSYSTEMS AND INTERFACES

The logical system given in the overview consists of a number of subsystems. Each subsystem addresses a basic function and together they achieve the system objectives. It is not possible to say how many subsystems should constitute a complete system since that is dependent on the system objectives. But a typical business application system should consist of four to six subsystems.

Any system serves three basic functions:

a. Input data into the system

b. Process entered data, as needed

c. Produce output from the system

Accordingly the subsystems of any system must provide the following basic capabilities:

a. Data entry and data edit

b. File update

c. Implementation of proper algorithms, decision tables, etc.

d. Report generation

The subsystems do not hang loose within the confines of a system. They are connected with one another so that a change in one subsystems generates ripple effects on the others. The interfaces among the subsystems accomplish that objective. An *interface* between two subsystems is a program that accepts the output from one subsystem and provides it as an input to the other. Let us now illustrate it with an example.

Suppose that for a given system, subsystem A accomplishes the data entry and edit process and subsystem B implements the file update process. Obviously, the file is updated with data that have passed all the edit routines. An interface between subsystems A and B accepts the clean edited data from subsystem A and provides them as input to subsystem B. As a result, files are updated with valid data. The edited data, which are output by subsystem A, may be stored in a temporary transaction file. This file may then be used by the interface to provide input data for subsystem B. If an edit routine in subsystem A is modified, the interface between subsystems A and B carries that information to subsystem B.

3.7   DATA FLOW DIAGRAMS

Davis ([1], p. 282) has described a data flow diagram as follows:

> A dataflow diagram is a logical model of a system. The model does not depend on hardware, software, data structure, or file organization: there are no physical implications in a dataflow diagram. Because the diagram is a graphic picture of the logical system, it tends to be easy for even nontechnical users to understand, and thus serves as an excellent communication tool. Finally, a dataflow diagram is a good starting point for system design.

There are five parts of a data flow diagram.

a. Source. The source is represented by a square or rectangle and signifies the starting point of the diagram.

b. Destination. Like the source, the destination is also represented by a square or a rectangle. It signifies the end point of the diagram and is also called *sink*.

The source and destination together constitute the boundary of the diagram. If the diagram represents a total system, the source and destination represent links of the system with other systems. If the diagram depicts a subsystem, the source and destination provide interfaces with other subsystems within the system. If a source or a destination appears more than once in a diagram, then its symbol, i.e., a square or a rectangle, is marked with a short diagonal line in one corner.

c. Process. A process is represented by a circle or an oval. It transforms data; i.e., the input to a process is always transformed into the output by application of the process. A process is not necessarily a program. It may represent a series of programs, a single program, or a module in a program. It may even represent a manual process such as keypunching or manual collection of source documents. Normally, in a data flow diagram of a total system a process comprises several programs. However, as the diagram is decomposed into finer parts, a process becomes closely analogous to a program.

d. Data flow. Data flow is represented by an arrow in the diagram and signifies data elements on a global level. For example, if a data flow diagram represents the data entry process, then a data flow may be labeled "Entered Data." It then consists of all the data elements or fields belonging to the form that has been entered. Thus, a single data flow normally represents a group of data elements or fields. It thus resembles the 01 level data in the Data Division of a COBOL program.

e. Data store. Data store is represented by a pair of horizontal parallel lines or by a three-sided (i.e. open-ended) rectangle. It is different from a file. It may represent a file, a part of a file, elements in a database, or even a portion of a record. Some or all of the data flows shown in a diagram should belong to the data stores included in that diagram.

Figure 3-1 shows the symbols used in a data flow diagram. Several assumptions and conventions are used in a data flow diagram:

1. The flow of data in the diagram starts at the source in the upper left corner and moves toward the destination in the lower right corner.
2. If two data flows entering into a process or coming out of a process are connected by the logical AND condition, then an asterisk (*) is placed between the two arrows representing the two data flows.
3. If two data flows entering into a process or coming out of a process are connected by the logical exclusive OR condition, i.e., EITHER–OR–BUT–NOT–BOTH condition, the symbol $\oplus$ is placed between the two arrows representing the two data flows.
4. Error processing or handling unusual condition is ignored in the diagram.
5. Housekeeping functions such as opening or closing files are not shown.

In summary, we can say that a data flow diagram merely shows *what* happens in the system or the subsystem but pays no attention to *how* it happens. As such it is the ideal graphic tool to describe a logical system. It thus differs from the system flowchart in that the latter is implementation bound, whereas the data flow diagram is not.

There are several levels of a data flow diagram. Usually, level 1 diagram represents the total system on a global or high-level basis. Successive levels, level 2, level 3, etc., of data flow diagrams show more and more details of

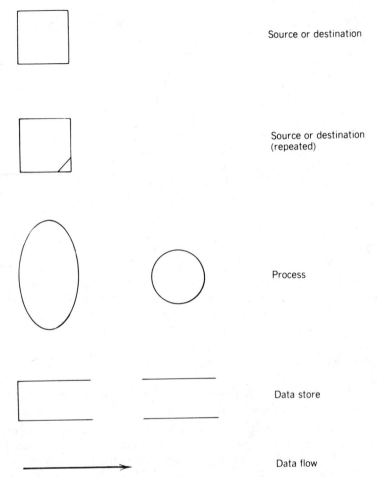

Source or destination

Source or destination
(repeated)

Process

Data store

Data flow

FIGURE 3-1　Symbols of the data flow diagram.

each subsystem. There is no strict rule regarding the level to which a data flow diagram should be broken down. But the third level appears adequate for most systems. Smaller or less complicated systems may even use only the first and second level diagrams.

3.8　EXAMPLE OF A DATA FLOW DIAGRAM

Let us consider a system consisting of the following three subsystems (see Section 3.6):

a. data entry and edit
b. file update
c. report generation

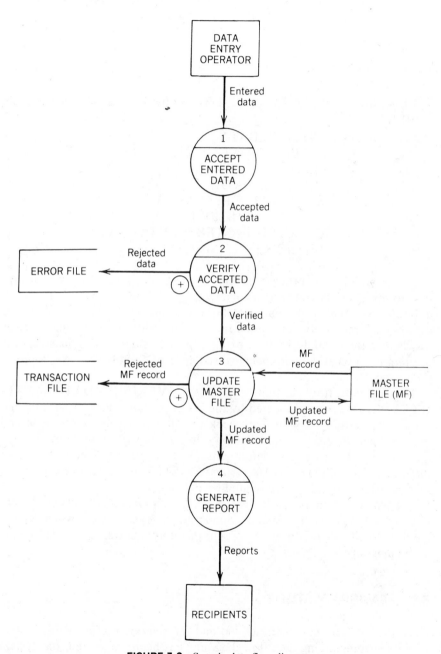

FIGURE 3-2 Sample data flow diagram.

Figure 3-2 shows the data flow diagram for level 1 for our system.
The diagram has the following components:

1 Source   DATA ENTRY OPERATOR

1 Destination   RECIPIENTS

4 Processes labeled 1, 2, 3, and 4

3 Data Stores   MASTER FILE, ERROR FILE, and TRANSACTION
FILE

8 Data Flows   ENTERED DATA
　　　　　　　ACCEPTED DATA
　　　　　　　VERIFIED DATA
　　　　　　　REJECTED DATA
　　　　　　　MASTER FILE RECORD
　　　　　　　UPDATED MASTER FILE RECORD
　　　　　　　REJECTED MASTER FILE RECORD
　　　　　　　REPORTS

The purpose of each process is self-explanatory. Since a process always
transforms a data flow, the name of an input data flow to a process is always
different from that of the output data flow to the same process. The three
data stores are merely conceptual files. No decision has yet been made as
to whether they should be tape or disk files. In fact, the emphasis of the
diagram is on what data are being stored and not on how they are stored or
processed.

The two data flows output by process 2, VERIFIED DATA and RE-
JECTED DATA, are connected by the exclusive OR relation, since the accepted data either pass all edit check or fail at least one edit check and
therefore get rejected. Hence two data flow symbol are related by the sign
$\oplus$. Similarly, in order to update the master file, process 3 requires both of
the data flows, VERIFIED DATA and MASTER FILE RECORD. Hence
their respective symbols are related by the sign *.

The whole system is started by the DATA ENTRY OPERATOR, which,
therefore, appears as the source. Similarly, the system ends when reports
are distributed to RECIPIENTS, which, therefore, appears as the destination
in the diagram.

3.9 FEASIBILITY STUDY

Once the problem has been defined and the overview of a logical system
with its accompanying data flow diagrams has been completed, the system
team examines the feasibility issues. The central question here is, Can the
system be implemented using the existing technology and available funds
within the client's organization?

If we analyze the above question, three separate feasibility issues appear:

a. Technical feasibility
b. Economic feasibility
c. Operational feasibility

3.9.1 Technical Feasibility

Technical feasibility involves the issue as to whether the necessary technology exists currently to implement the system. It is a relevant question for systems involving weaponry and military strategy. The so-called "Star Wars" defense system is still not technically feasible. A computer that can think and make decisions like human beings is technically not feasible. Consequently, a system that assumes the availability of such computers is also not feasible from a technical standpoint. The situation is, however, very different in the business or standard scientific and engineering environment. Due to the advanced nature of the hardware and software technology in the mid-1980s, practically any business or standard scientific application system can be implemented and hence is technically feasible.

3.9.2 Economic Feasibility

Economic feasibility is usually the most stringent restriction imposed on the study, because the company cannot allocate an unlimited amount of funds to the project. The maximum available amount of funds usually restricts the scope of the project. Therefore, after deciding that the system is technically feasible, the system team makes an estimate of the total cost to design and implement the system. Of course, at this early stage the cost estimate is bound to be inaccurate. A range of ±50 percent over the estimated cost figure is acceptable. If the upper limit of the estimated cost, i.e., the estimated amount +50 percent, falls within the allocated budget, the system is declared as economically feasible. The economic feasibility is closely related to the issue of cost/benefit analysis of the system. The user will always want to justify the expenses incurred in the design and implementation of the proposed system by weighing the cost against the potential benefits of the system. We shall discuss methods of doing cost/benefit analysis in Chapter 5.

3.9.3 Operational Feasibility

Operational feasibility is related to the question of whether the proposed system can be implemented in the user's organization. For example, if the proposed logical system requires a decentralized database while the company policy mandates that all data must be stored at the company headquarters, then the system faces operational infeasibility. Similarly, if the computer system in the user's organization is obsolete or overloaded, then the system team may suggest that an outside data processing service facility be used. But if the company's data are highly classified or sensitive to security issues, then such implementation becomes operationally infeasible. The system team may propose necessary modifications to the existing operations if it feels that the operational feasibility is lacking.

The amount of time to be devoted to the problem definition and feasibility study depends on the scope of the system. If the system involves only a minor modification of an existing system, then a day or two would be sufficient. If it involves the design of a new invoice system, then the feasibility study should extend over 2 or 3 weeks. If the user wants a system to address a totally new problem costing several million dollars, then perhaps 6 to 8 weeks would be needed. In such cases, it may even be advisable to design and implement a prototype system at first. If the prototype works well, then it may later be made a regular operational system. Otherwise, the system team learns from the deficiencies or shortcomings of the prototype, modifies the prototype so as to eliminate the defect, and implements a better and tested system. As a rule of thumb, the cost of completing the problem definition and feasibility study should be approximately 5 to 10 percent of the estimated total project cost.

3.10   END PRODUCT OF FIRST PHASE

Problem definition and feasibility study comprise the first phase of the system development. On completion of this phase, the system team prepares a report documenting all findings and outlining their recommendations. This report is aimed primarily at upper level management of the user's organization. Hence the body of the report should avoid too many technical details. Ideally, the main report should contain findings and recommendations, whereas supporting technical details should be included as appendices or as a second volume. If the appendices become as bulky as the main report, a second volume should be planned.

We now describe the contents of the report concluding the first phase of study.

3.10.1   Background/History/Introduction

Describe the user's organization, including the organization chart. Give some history on the development of the company in order to provide a background for the problem description. Then specify the current problem and show how its impact on the organization. The definition of the problem should be clear and concise and limited to one page. Then discuss how the team arrived at the problem definition. For example, describe the interview process, review of current operating procedures, and review of earlier studies on the subject.

If necessary, the detailed proceedings of all the interviews may be included in an appendix. Likewise, titles of current manuals and earlier studies reviewed can be provided as a reference section in an appendix.

3.10.2   Objectives of the Study

Describe what the team wants to do in order to address the problem. Discuss the functional capabilities of the proposed system. Identify them in the form of a list rather than as detailed descriptions, because it is much simpler to scan a list than a narrative. If any of the capabilities is introduced to provide a future enhancement as opposed to addressing the immediate problem, mention that fact. This will give the user a chance to assign priorities to the list of capabilities. But always restrict the capabilities to the issue of *what* should be done for solving the problem, not *how* that can be done.

3.10.3   Description of the Proposed System

Describe the proposed logical system that will provide all the capabilities listed in Section 3.10.2. Then describe the subsystems and their interfaces comprising the complete system. It is not necessarily true that each capability should be implemented via a different subsystem. However, when describing the subsystems and their interfaces, correlate each subsystem with one or more functional capabilities that it is going to implement. This section should be designed in the form of structured narrative. Clearly describe each subsystem and its interfaces by indicating the input and the output for the subsystem.

3.10.4   High-Level Data Flow Diagram

Design one data flow diagram at level 1, i.e., on the global level. It comprises the entire system. Describe each subsystem as one or more processes and the interfaces among the subsystems as appropriate data flows among the processes. The boundaries of the system are depicted as sources and destinations. Number each process as an integer, 1, 2, 3, . . . etc. This makes it easier to reference future expansions of any given process; For example, process 2 can be further decomposed into subprocesses labeled 2.1, 2.2, 2.3, . . . etc.

3.10.5   Feasibility Study

Discuss the three feasibility issues, namely, technical, economic, and operational feasibilities. Address each issue separately and justify your conclusions. If the allocated budget is insufficient for implementing the total system, say so and indicate how much extra money is needed to develop the complete system and/or what parts of the system can be implemented within the allocated funds. If there is operational infeasibility, describe what changes are needed in the organization in order to implement the system. Always document your findings and recommendations. Remember that the continuation of the work depends heavily on the feasibility study.

3.10.6 Executive Summary

At the beginning of the report include a summary of the problem definition and the key findings. This gives top management an overview of the system study.

3.11 MANAGEMENT REVIEW

After submitting the phase 1 report to the users for their review, the system team must make a presentation of the key findings. This is essentially a sales pitch and works as a good marketing technique. At this initial level, the presentation is aimed at top management so as to enable them to authorize further works and allocate funds to continue such works. Since the presentation is geared toward upper level managers in the user organization, technical details should be kept to a minimum. The system team should clearly formulate the problem and the proposed logical system. If there are any feasibility problems, they should be mentioned because the top management would always want to know the viability of the solution in light of technical, economic, and operational considerations.

As a result of the presentation, management should be able to determine that

a. The system team really understands the problem.
b. The system team has clearly defined what should be done to solve the problem.
c. The system team has examined the three feasibility issues adequately.

The management review process may be iterative. If there are questions raised that cannot be answered right away, the system team may ask for additional time. Then a follow-up presentation is arranged at a later date, and the necessary clarifications are made. At any rate, the final goal of the management review is to gain authorization for future work.

3.12 SUMMARY

This chapter discussed the first phase of the system life cycle, namely, problem definition and feasibility. This is an iterative process. The system team has to redefine and expand the problem definition a number of times before finalizing it. Often the customer may tell about the symptoms without identifying the problem. Two techniques are widely used to gather facts that enable the system team to define the problem:

a. Interview different levels of personnel involved with the problem.
b. Review current operations manuals and any available reports or studies done earlier in order to address the problem.

These two processes may sometimes give conflicting pictures. The interviews may indicate that certain operations are done in one way, whereas the operations manuals may show that they should be done differently. In such cases, the team must resolve the conflict before formulating a clear-cut definition of the problem.

The system analyst who conducts the interviews should have excellent communications skill. The interview starts with the top level management and supervisory personnel and gradually moves to the lower level personnel as more and more details are needed. Any interview session has three stages:

a. **Interview preparation.** The interviewer makes a list of interviewees in accordance with the objectives. In addition, he or she prepares a list of questions to be asked or discussed during the interview. This list must remain flexible since the progress of the interview cannot be anticipated fully in advance. Finally, the interviewer notifies the interviewee(s) of the time, place, and purpose of the interview by means of written memos.

b. **Interview session.** The actual interview starts with the interviewer identifying him- or herself and diplomatically pointing out that the top management has authorized the system study. The interviewer then follows the list of questions and takes notes as needed. If the interviewee gives prior consent, the interviewer may use a tape recorder to tape the session. However, during the interview he or she should listen more than talk and should never appear to be confrontational. At the end of the session, the interviewer should promise a written follow-up summarizing the transactions during the interview.

c. **Interview follow-up.** The interviewer prepares a written report summarizing the main findings of the interview. This enables the interviewee to clarify any misunderstanding and also establishes good documentations for future reference. If necessary, the interviewer may schedule a follow-up interview session.

Besides the interviews, the system team reviews current operating procedures and reports of earlier studies, if any. Such documents provide additional insight into the problem definition.

Finally, the team is able to give an overview of the proposed logical system. This overview contains, as a minimum, the following items:

a. Problem statement
b. Objectives of the system
c. System boundaries
d. A tentative solution
e. Project cost

The last item is only an approximate cost and can vary within a range of ±50 percent. The system overview should describe the major subsystems of the system and should specify the interfaces among the subsystems.

A data flow diagram is used as the graphic tool to describe the subsystems and their interfaces. The diagram can show details at multiple levels. Any such diagram consists of five components:

a. Source. Signifies the starting point of the system
b. Destination. Signifies the end point of the system
c. Process. Indicates how data are transformed within the system
d. Data flow. Represents data elements, usually on a global level
e. Data store. Represents the repository of data flows

The data flow diagram merely shows *what* happens in the system, but not *how* it happens. It thus is independent of implementation and differs from a system flowchart, which is implementation bound.

The final stage of the first phase consists of the feasibility study. Three issues are discussed during this stage:

a. Technical feasibility. Does the necessary technology exist to implement the logical system? The answer is normally in the affirmative as far as ordinary business systems are concerned.
b. Economic feasibility. Can the logical system be implemented within the allocated budget? This is usually the most stringent constraint imposed on the project, since almost any system can be implemented if unlimited funds are available.
c. Operational feasibility. Can the proposed system be implemented within the customer's organization?

At the end of the first phase of the system study, the system team prepares an extensive report documenting all findings and outlining their recommendations. This report is aimed primarily at upper management and is, therefore, less technical in content. Normally, it should have the following sections:

a. Background/history/introduction
b. Objectives of the system study
c. Description of the proposed system
d. High-level data flow diagram
e. Feasibility study
f. Executive summary

The last section usually appears at the beginning of the report and includes a summary of the problem definition and major findings.

After submitting the report, the team makes a presentation of its key findings and recommendations and invites the audience (usually upper level management) to ask questions or make comments. This gives the management a way to clarify any confusion. If management agrees that the system team has understood the problem and has clearly outlined what should be done to address the problem, permission is granted to conduct the second phase, i.e., system analysis, of the system study.

3.13   KEY WORDS

The following key words are used in this chapter:

data flow	levels of data flow diagram
data flow diagram	logical AND
data store	logical system
destination	operational feasibility
economic feasibility	operations manual
exclusive OR	problem definition
feasibility study	process
interfaces	project cost
interview follow-up	sink
interview preparation	source
interview session	subsystem
interviewee	system boundaries
interviewing techniques	technical feasibility

REFERENCES

1. William S. Davis, *System Analysis and Design: A Structured Approach*, Addison-Wesley, Reading, MA, 1983.
2. Henry C. Lucas, Jr., *The Analysis, Design, and Implementation of Information Systems*, McGraw Hill, New York, 1985.

REVIEW QUESTIONS

1. Why is the problem definition regarded as the starting point of a system life cycle?

2. What role does an organization chart play in defining the problem?

3. Explain why the interview process is iterative in nature.

4. Why should the analyst start with interviewing upper-level employees and then descend to the lower level employees?

5. Assume you are interviewing the supervisor of customer services of a company in order to gather information about the monthly billing procedure followed by the company. You have been told that the company receives frequent complaints from the customers about inaccurate monthly bills. Plan an interview session and prepare the following:

   a. A memo to the interviewee outlining the purpose, time, and place of interview

   b. A list of tentative questions in priority order

   c. A report containing the transactions of the interview

6. Can you get conflicting information from interview and review of operations manual? If so, how can you resolve the conflict?

7. Define subsystems and interfaces within a system.

8. What do you mean by system boundaries?

9. What is a data flow diagram? How does it differ from a system flowchart? Why is it specially suitable to describe a logical system?

10. What are logical AND and exclusive OR operations? How are they represented in a data flow diagram?

11. What is meant by a feasibility study?

12. Explain the three feasibility issues, namely, technical, economic, and operational feasibility.

13. What items should be included in the report describing the outcome of the problem definition and feasibility study?

14. What is the purpose of having management review the end product of a problem definition and feasibility study?

Problem Definition and Feasibility: Case Studies

4.1 INTRODUCTION

In this chapter we shall discuss the first phase of system development for two case studies: an order processing system and a financial reporting system. For each case, we shall prepare the end product as a report consisting of the following parts:

- **a.** Background/History/Introduction
- **b.** Objectives of the Study
- **c.** Description of the Proposed System
- **d.** Level 1 Data Flow Diagram
- **e.** Feasibility Study
- **f.** Management Review

4.2 ORDER PROCESSING SYSTEM: BACKGROUND/HISTORY/ INTRODUCTION

Toy World, Inc., was established in 1965 in Harrisburg, Pennsylvania, as a small toy shop. Initially it was a family-run business. As a result, there was little management overhead and customers got largely discounted prices, leading to a quick expansion of Toy World's business. By 1975 it was necessary to open three branch stores in Philadelphia, Scranton, and Wilkes-

Barre, Pennsylvania. The original Harrisburg store operated as the headquarters.

The lead time required for design, manufacturing, and final distribution of the toys at the retail level was 15 months on the average. Since the toy business is about 60 percent seasonal in nature, it is necessary to keep track of inventory and place orders promptly to meet the seasonal demands. During a typical year, the highest volume of sales occurs during the Christmas season starting in mid-November and continuing through Christmas eve. Two other occasions of increased demand occur during Easter and Halloween. The order processing system has to keep track of all these variable demand volumes and replenish the inventory accordingly.

At present, Toy World has a president, three vice presidents—one at each branch store—and about 200 other supporting staff. Figure 4-1 shows the organization chart of Toy World. There are 70 salespeople in the four shops. Toy World's data processing department has a VAX-11/785, which is in operation about 300 hours a month. Its principal applications include payroll, billing, accounts receivable, preparation of invoices, and inventory control. The entire operation is centralized at the Harrisburg location, which employs one MIS manager, two application programmers, one system programmer, one operator, and one data entry clerk. The MIS department is aggressively looking for automating new applications. The order processing system is the first item on its list.

Currently the order processing system operates as follows: A salesperson fills an order form for a customer and enters the item numbers and prices for the requested toys. He or she also indicates whether substitutions are permitted. The order form is then sent to the Harrisburg headquarters. There the staff checks if the items are available in the inventory. If so, they are shipped to the customer. Otherwise, an order is placed with the original manufacturer of the items. For seasonal demands, however, this cumbersome manual process takes too much time and results in customer dissatisfaction and even customer loss. In addition, Toy World does not make any sophisticated forecasts and, as a result, does not get all its inventory updated before the start of each season of peak demand. Thus, during the last 3 years, Toy World has suffered heavy losses at Christmas.

The president of Toy World recently discussed the issue with the three vice presidents and the MIS manager. They decided that a system consultant should be invited to address the problem. An invitation was sent to the firm, High Price Consultants (HPC for short), which did a satisfactory job for Toy World in the past. Two staff members of HPC came to Harrisburg to investigate the situation. They decided to make a further study; see Figure 4-2 for their response memo.

4.2.1 Preliminary System Study and Problem Definition

After getting the authorization from Mr. Dennis Friend, the president of Toy World, representatives from HPC spent 3 days at Harrisburg and 1 day at

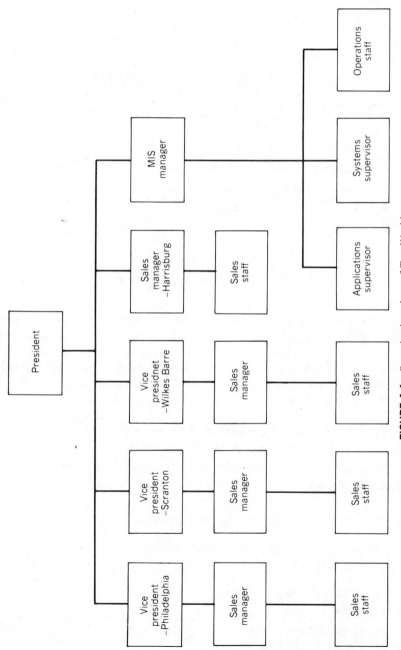

FIGURE 4-1 Organization chart of Toy World.

57

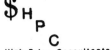

MEMORANDUM

High Price Consultants
Boston, Mass

TO: Mr. Dennis Friend

DATE: 17 SEPTEMBER 1983
FROM: HIGH PRICE CONSULTANTS
PHONE: 999-9991
PROJ #: 8211

SUBJECT: TOY WORLD, INC., PA.

In regards to our meeting on September 5, we (HPC) have evaluated the new system. As a result of the tour and several meetings with yourself, a proposal is being developed.

I estimate the documentation shall take approximatly 14 working days to design. If you should have any questions, please feel free to call.

Regards.

P.S.   Thank you for your generosity at the luncheon.

FIGURE 4-2   Memo from HPC, Inc.

each of the Philadelphia, Scranton, and Wilkes-Barre branches. Through a series of interviews and a review of manuals and operating procedures of order processing, invoicing, and inventory management, HPC prepared a system proposal for Toy World. According to the proposal, the following problems appeared to be causing the customer dissatisfaction and loss of business for Toy World:

a. Lack of motivation and control of the sales force
b. Inability to forecast market trends accurately
c. Unavailability of an updated finished goods inventory file
d. Lack of a well-defined policy for filling customer orders with substitutions
e. Lack of control over production schedules

Accordingly, HPC recommended that the proposed system study be focussed on the following areas:

1. Order Entry System
2. Inventory Control
3. Production/Operations
4. Marketing Applications
5. Sales Order/Transaction Processing
6. Sales Analysis

HPC summarized their recommendations as follows:

It is recommended that a system study be initiated to investigate how the above basic business applications can be fully used to improve Toy World's organization. The system in use appears to consist of an independent manual and computerized processing applications without any framework for organizational or relational dependency. It is possible that a management information system (MIS) could not only process data generated by the business's operations but also assist in the support of management decision making.

4.2.2 Objectives of the Study

As far as the objectives of a system study go, the current system should be looked at to see if it could handle the added load of doing more than just routine office processing. Toy World needs to use the current system, upgrade it if possible, get a larger system that can handle the added workload or possibly get a second separate system to use as a market-analysis tool, improve service on the accounts, reduce costs through better control over inventory and other expenses and production and production-control, and get more information on marketing, sales research, and accounting, for a start.

Also, the way in which the order entry system works needs to be evaluated and most likely changed from the older card-type system to a modern and accurate on-line terminal system, using the inventory file to eliminate ordering more of an item than is actually on hand.

Thus, the proposed system will provide Toy World with an automated order-entry system, including a finished goods inventory file and a sales information file. The following functional capabilities will be made available:

1. The proposed system will handle backorders, eliminating the manual process of retyping invoices on partially filled orders.
2. The proposed system will supply management with up-to-date information on sales and inventory.
3. The proposed system will handle customer requested changes and cancellations. The user will be able to decide upon
   a. Automated substitution procedures
   b. Sequencing of "price-title" categories and carryovers
   c. Future selling trends
4. Future applications can be developed, if desired, to supply management with sales agent information.

4.2.3 Description of the Proposed System

The proposed logical system will have three major components:

   a. Order processing
   b. Inventory control
   c. Sales management

The order processing subsystem will be triggered by the sales transactions and will interface with the inventory subsystem via inventory data and customer data. It will also interface with the sales subsystem via sales data. The philosophy behind the logical system is that the flow of data is initiated by means of a sales transaction, either as a direct sale to a customer or as a customer requested change to an earlier sale transaction. Therefore, the sale of an item or sales change data provides the interface between the order processing subsystem and the sales subsystem. After an item is sold, the inventory level for that item is reduced. This data provides the interface between the order processing subsystem and the inventory control subsystem. Eventually, all three subsystems jointly contribute toward generation of reports for various levels of management.

4.2.4   Level 1 Data Flow Diagram

Figure 4-3 is the level 1 data flow diagram for the complete system. It consists of six processes distributed over three subsystems as follows:

Subsystem	Process Numbers
Order Processing	1, 4, and 5
Inventory Control	2 and 6
Sales Information	3

The system has two sources—SALES and CUSTOMER SERVICE—and two destinations—WAREHOUSE and MANAGEMENT. Together, they determine the boundaries of the system. An Order Transaction from SALES initiates process 1 (Process Order), and a change Transaction from CUSTOMER SERVICE initiates process 4 (Process Customer Changes). Both of these processes access the two data stores

```
D2 : MASTER FILE
D4 : CUSTOMER FILE
```

and update them by incorporating the necessary changes. The updated MASTER FILE data then feeds into process 5 (Process Master File), which generates a Pickup List and a Detailed Invoice and sends them to the WAREHOUSE. Thus, processes 1, 4, and 5 together comprise the Order Processing Subsystem.

The three data flows, Inventory Data from process 1, Inventory Change Data from process 4, and Change Data from process 2, feed into the data store

```
D1 : INVENTORY,
```

which then sends the data flow, Inventory Status, to process 6 (Generate Inventory Reports). This latter process produces Inventory Status Report

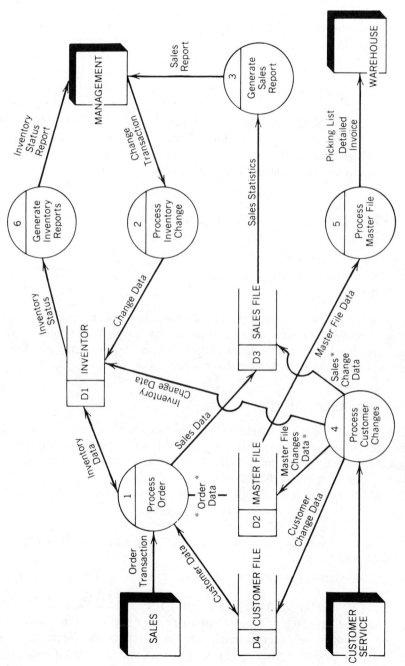

FIGURE 4-3 Data flow diagram for order processing system.

for MANAGEMENT. If necessary, MANAGEMENT may send Change Transaction to process 2 in order to make changes in the data store D1. Processes 2 and 6 together constitute the inventory control subsystem, which keeps the inventory up to date and provides appropriate status reports for management. Process 3 (Generate Sales Report) takes care of the sales information subsystem. It accesses the data store

    D3 : SALES FILE

in order to gather Sales Statistics for generating necessary Sales Report for MANAGEMENT. The data store SALES FILE gets two data flows, Sales Data and Sales Change Data, from processes 1 and 4, respectively.
    The interfaces among different processes are as follows:

Interfaces (Data Flows)	between Processes
Order Data, Masterfile Change Data	1, 4
Inventory Data, Inventory Status	1, 6
Sales Change Data, Sales Statistics	4, 3
Order Data, Masterfile Data	1, 5
Change Data, Inventory Status	2, 6

As a result of these interfaces, the six processes are interconnected as parts of the complete logical system. At the design and implementation stages, these interfaces are coded as data elements that are passed among different program modules that implement the six processes.

4.2.5 Feasibility Study

HPC examined the three feasibility issues—technical, economic, and operational. Since the system will use standard computer applications, the technical feasibility posed no problem. Next, HPC investigated the economic feasibility and concluded the following:
    Toy World management has identified the following two major economic constraints to be considered in selecting a system:

a. A new system should require an initial investment of not more than $400,000.

b. The maximum payback period for the system should be 3 years.

1. A NEW SYSTEM SHOULD COST LESS THAN $400,000. The investment department cannot authorize an expenditure of more than $400,000 for computer systems without formal board approval.

2. A NEW SYSTEM SHOULD HAVE A PAYBACK PERIOD OF LESS THAN 3 YEARS. It is company policy that all investments in equip-

ment or software for new information systems have a maximum payback period of less than 3 years, because the company can achieve this rate of return from other investments of similar risk.

Finally, the operational feasibility was considered. Toy World insisted that the computer system and all the data files be located at its Harrisburg headquarters. HPC did not regard that as any problem. In fact, since the three branches were rather small, each one separately might not justify a satellite computer system locally.

4.2.6 Management Review

HPC arranged a formal presentation of its findings for the management of Toy World, Inc. The three vice presidents from the three branches, along with the upper level management at the Harrisburg headquarters, attended the presentation. The session was well received, and Mr. Dennis Friend, the president of the Toy World, asked HPC to continue its investigations into the next phase—system analysis. Figure 4-4 shows the letter of appreciation and acknowledgment that HPC wrote to Mr. Friend.

HPC, Inc., is now ready to venture into the detailed system analysis phase.

4.3 FINANCIAL REPORTING SYSTEM

The Massachusetts Educational Foundation (MEF) is a private nonprofit organization located in Boston. MEF awards grants to educational institutions nationwide for research activities. In return, each grantee, provides MEF with data on

- ☐ Assets, liabilities, and capital
- ☐ Revenue items
- ☐ Expense items
- ☐ Nonfinancial operating expenses

MEF receives such data continuously during the year, enters them into their computer system, and produces routine periodic reports. At the end of each year, MEF prepares a comprehensive annual report containing over 50 statistical tables most of which are multipage. The process of data collection and data verification are done centrally in a simple data processing facility that houses the computer and is located in Boston.

MEF started its operation in 1978. Its financial reporting system was then completely batch oriented. Data were input via keypunched cards. The amount of data editing was minimal. In addition, the cross-checking of input data across forms submitted by the grantees could not be done easily due

October 7, 1983

$H
 P
  C

High Price Consultants

Boston, Mass

Mr. Dennis Friend, President
Toy World, Inc.
Harrisburg, PA 17112

Dear Mr. Friend:

HPC welcomes the opportunity to provide Toy World organization
with effective improvements to your business activities. To meet
your requirements, HPC is proposing to implement a new Management
Information System (MIS). The proposed system will resolve your
present problems which exist in the following areas:

1. Inventory Control
2. Order Entry
3. Production Operations
4. Marketing Applications
5. Sales Order Transactions
6. Sales Analysis

In order to provide major solutions to your present situation, HPC
plans to enter upon a further investigation. This in depth
investigation will provide your oganization with a MIS system that
will manifest itself not only in customer services and quality
services, but improvement in sales, profit, management decision
marketing, a faster and better response to actions of competitions,
and, especially, complete and thorough monitoring of your
inventory status.

HPC firmly believes that with the development of a computerbased
information system, you can achieve complete control of your Toy
World organization.

If you should have any questions regarding this major innovation,
please do not hesistate to call upon HPC at your earliest
convenience.

Very truly yours,

High Price Consultants

FIGURE 4-4 Acknowledgment letter from HPC.

to the batch characteristic of the system. The data analysts wanted to have
an on-line query capability that was totally absent in the batch system. As
a result, there was an average time lag of 6 weeks from the time the data
first entered the system and the time it ultimately became clean. This often
led to delays in producing reports.

In early 1984, MEF contacted the system consulting team HPC (see Section 4.2) to convert the batch system into an on-line system. MEF specified
the following requirements as a minimum:

a. The data entry operator should be able to enter data on-line interactively.

b. The data will be subjected to routine edit checks with appropriate
prompts and error messages.

c. The edited data will then be subjected to cross-checking of values across multiple forms.

d. The finally validated data will be entered into a database.

e. A user-friendly DBMS package will be used to query the database and generate routine periodic reports as well as ad hoc query reports.

f. The data in the database will ultimately be used to produce the annual statistical report.

4.3.1 Objectives of the Proposed System

The system proposed by HPC is called *Financial Reporting System* (FRS for short). It aims at the following objectives:

a. Data entry/edit/display
b. Data validation
c. Database update/retrieval
d. Database status recording

Input to FRS is provided through 10 forms, some of which are multipage.

(a) Data Entry/Edit Display. The user can perform the following functions:

i. Call up a specified blank form on the CRT screen
ii. Enter data interactively into the displayed form
iii. Edit the entered data
iv. Respond to error messages and prompts displayed on the top of the screen
v. Store each page for subsequent retrieval after data entry is finished
vi. Display different pages of a multipage form

(b) Data Validation. The purpose of data validation is to ensure that the data entered into FRS is valid. The following types of checks are provided:

i. **Arithmetic check.** Checks a row total or a column total to verify that the number entered as total equals the sum of the component row or column items.

ii. **Consistency check.** Checks if one data value equals another data value on the same form or on different forms.

iii. **Range check.** Checks whether a data value falls between a minimum and a maximum value already established.

iv. **Relational checks.** Checks whether a certain preestablished equation or inequality involving two or more data values is satisfied. This is really an extension of type (i).

v. Alphanumeric validity check. Checks whether a designated field matches an entry in a preestablished table of valid values.

(c) Database Update/Retrieval. The database update/retrieval functions enable the user to do the following:

i. Modify and store form data into an appropriate form relation in the forms database

ii. Use relational DBMS commands to view the contents of any form relation

iii. Generate ad hoc reports from one or more form relations by using relational DBMS commands

(d) Database Status Recording. The database status recording function is provided by means of a relation called database status relation in the forms database. The contents of this relation answer the following questions:

i. Is a particular Form/Page contained in the database?

ii. When was it

    a. last updated?

    b. originally entered?

    c. last processed by the validation subsystem?

iii. What is the validation status?

The answer to the last question is provided by means of a status flag, which can take the following values:

```
00   Not validated.
01   Processed by validation routine that found
     at least one problem.
02   Processed by validation routine that did
     not find any problems.
```

Whenever a form is entered or updated, the validation flag should be set to 00.

4.3.2 Operational Scenarios

In order to help the customer understand how the proposed FRS will work, HPC prepared the following operational scenarios:

An authorized user can access the system via a front end which is called Executive Module (EM). The user can login either as USER or ADMIN. The login name USER provides a version of EM offering the following options:

1. Data Entry/Edit/Display
2. Data Validation
3. Exit System

The login name ADMIN provides the following four options:

1. Data Entry/Edit/Display
2. Data Validation
3. Database Administration
4. Exit System

ADMIN version of EM has a higher level of security and is intended only for system managers and other similar personnel. USER version of EM, on the other hand, is to be used by data entry operators or validation analysts for entering data and/or for generating hardcopy forms.

Options 1 and 2 under USER or ADMIN correspond respectively to items (a) and (b) of Section 4.10. Option 3 under ADMIN represents items (c) and (d) together of Section 4.10.

Under FRS, a form is entered initially through an interactive screen and is stored in a temporary file. At the end of each day, all the forms entered during the last 24 hours are transferred to the corresponding form relations in the forms database, which is the permanent repository of the form data. In order to edit the form, the user brings it from the database to a temporary storage file and incorporates all changes; the modified form is then sent back to the forms database at the end of the day as a batch job.

4.3.3 Description of the Proposed System

FRS consists of four major subsystems:

a. Data Entry/Edit/Display
b. Data Validation
c. Database Administration
d. Annual Report Generation.

Subsystem (a) implements the functional capabilities (i)–(vi) of item (a) in Section 4.3.1. After logging on either as USER or ADMIN, the user enters his or her option of Data Entry/Edit/Display subsystem and is prompted for four parameters:

☐ Fiscal year
☐ Grantee ID number
☐ Form number
☐ Category number (which signifies the grant category such as medical, scientific, etc.)

The parameters are then displayed for user review and can be changed if desired. Once the user confirms all four parameters, control passes to the subsystem monitor and a copy of the desired form is displayed on the screen. If no prior data were entered into the form, the copy appears as blank and the user enters data into the form. Otherwise, the form appears with the most recently entered data, and the user can review and/or edit the data. The cursor moves from one field to the next. If the entered data fails on edit check, an appropriate message or prompt appears at the bottom of the screen. In case of a multipage form, the next page automatically appears as soon as the current page is finished. After each page is complete, it is entered and saved in a data store called ENTERED FORM.

Subsystem (b) implements the five categories of checks labeled (i)–(v) in item (b) of Section 4.3.1. The user accesses this subsystem either as USER or ADMIN by selecting the option Data Validation on the menu. He or she is prompted for the four parameters:

- ☐ Fiscal year
- ☐ Grantee ID
- ☐ Form number
- ☐ Category number

As in subsystem (a), after the user confirms the entered values of the parameters, control passes to the subsystem monitor. It involves the appropriate checks, applies them to the form, and displays error messages, if any, at the bottom of the screen.

Subsystem (c) implements the functional capabilities (i)–(iii) of item (c) and item (d) of Section 4.3.1. The user can access this subsystem only by logging on as ADMIN. As in subsystems (a) and (b), the user enters and confirms the four parameters:

- ☐ Fiscal year
- ☐ Grantee ID
- ☐ Form number
- ☐ Category number

Then the control passes to the subsystem monitor, which displays the following options in a menu form:

1. Store form data in the relational database.
2. Display the contents of the form relation.
3. Generate ad hoc reports from the database by using DBMS query commands.
4. Check the update or validation status of the form relation.
5. Exit to the main menu EM.

The user types in the appropriate number to invoke the desired option.

Subsystem (d) is run only once a year and is not available as a menu item. Throughout the year MEF collects data reporting forms from all its grantees and stores them, after having them validated, in the forms database of FRS. At the end of each calendar year, the system accesses all the data and runs the report-generation subsystem to generate the annual report.

Subsystem (a) interfaces with subsystem (b) via the data flow, Entered/Edited Form. Subsystem (b) interfaces with subsystem (c) via the data flow, Validated Form. No direct interface exists between subsystems (a) and (c) because HPC has recommended that a form will not be stored in the forms database until it is completely validated and the database status flag has the value 02 (see Section 4.3.1.4).

4.3.4 Level 1 Data Flow Diagram

Figure 4-5 represents the level 1 data flow diagram of FRS. The system consists of five processes:

1. Enter/Edit/Display Data
2. Update Form Data Store
3. Validate Entered and Edited Form
4. Query and Maintain Form Database
5. Generate Annual Report

There are four data stores in FRS:

```
D1:  ENTERED AND EDITED FORMS
D2:  REJECTED FORMS
D3:  VALIDATED FORMS
D4:  NON-VALIDATED FORMS
```

USER appears as both the source and the destination for the diagram.

The system starts with the USER entering a Reporting Form into process 1. The data undergoes edit checks. If it passes all the checks, it is stored in D1, otherwise it is rejected and is stored in D2. The output from process 1 is the data flow Entered/Edited Form, which is input to process 2 in order to update the contents of the data store D1. After the form data is updated by process 2, it is sent to process 3 for validation. Once again, two possibilities arise: either the form is completely validated and stored in D3, or it fails at least one validity check and is stored in D4.

The validated forms in D3 are now sent to process 4 in order to store them as Form Relations in the Form Database. Process 4 enables the USER to access any such form and generate ad hoc query reports from the database. Finally, process 5 accesses the form relations as output from process 4 in order to generate the annual report. Of course, process 5 is run only once a year.

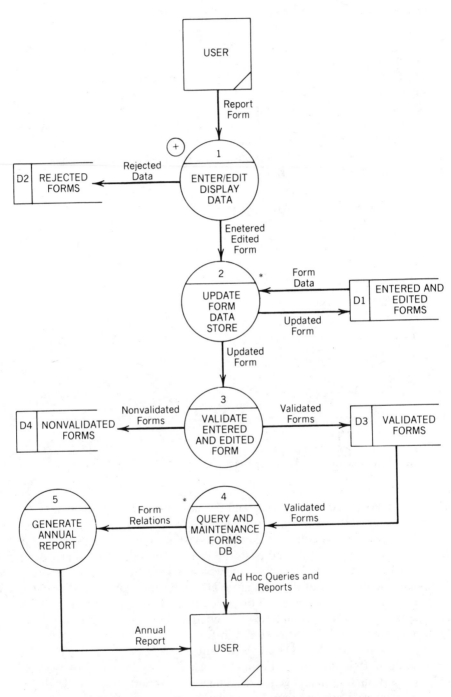

FIGURE 4-5 Data flow diagram for financial reporting system.

4.3.5 Feasibility of FRS

FRS is essentially a partially on-line transaction processing system with a supporting database for ad hoc query. Consequently, the current hardware and software technology are sufficient to implement the system. Thus the technical feasibility is assumed.

Based on a preliminary estimate, the consulting firm HPC thinks that it will cost nearly $250,000 to design and implement the system. Since the complete details of FRS are yet to be developed, the estimated cost may be overrun by as much as 50 percent; but since the customer MEF already owns the hardware, the upper limit of the possible cost overrun may be less. At a board meeting, MEF authorized an initial budget of $100,000 for the system study. The president of MEF expects to allocate an additional $200,000 for the project during the next 6 months. This indicates that the system is financially feasible.

Finally, the operational feasibility poses no problem. All the data are received and processed at the Boston headquarters of the company. The computer equipment and the data processing staffs are located there. Therefore, the data analysts can enter data, analyze them, and generate ad hoc reports from the database without problems. Also, they can change or reenter faulty data on-line.

On March 29, 1984, HPC arranged an oral presentation of the problem definition and feasibility study for the customer. The session was very well-received. The analysts especially liked the idea of being able to enter and edit data interactively. Since the present system is completely batch processing, changing already entered data creates a bottleneck and is unusually slow and frustrating. So, HPC finally got the approval for proceeding to the second phase of the project, i.e., the detailed system analysis phase.

4.4 SUMMARY

This chapter discussed two case studies in order to illustrate the techniques of problem definition and feasibility issues. The first case study involved an order processing system, and the second one involved a financial reporting system.

Toy World, Inc., was established in 1965 in Harrisburg, Pennsylvania, as a small toy shop. By 1975 it grew into a toy manufacturing and selling firm with branches in three cities in Pennsylvania. Partially due to its growth, Toy World has not been able to keep up with its sales and order processing systems. As a result, the company president requested that a system consulting company, High Price Consultants (HPC), conduct an initial system study to pinpoint the problem and propose a solution. After some preliminary investigations, HPC recommended that the study be focussed on the following areas:

1. Order Entry System
2. Inventory Control
3. Production/Operations
4. Marketing Applications
5. Sales Order/Transaction Processing
6. Sales Analysis

The proposed logical system will consist of three major components:

a. Order Processing
b. Inventory Control
c. Sales Management

The order processing subsystem will be triggered by the sales transaction and will interface with the inventory control subsystem via inventory data and customer data. Finally, all three subsystems jointly contribute toward generation of reports for various levels of management. Figure 4-3 is the level 1 data flow diagram for the system.

The proposed system is both technically and operationally feasible. Toy World management imposed two major financial constraints:

☐ The new system should not require more than $400,000 as the initial investment.
☐ The maximum payback period should be 3 years.

HPC feels that both of these restrictions can be met, so the project is economically feasible.

The Massachusetts Educational Foundation (MEF) is a private nonprofit organization located in Boston. MEF awards grants to educational institutions nationwide for research activities. In return, each grantee provides MEF with data on

☐ Assets, liabilities, and capital
☐ Revenue items
☐ Expense items
☐ Nonfinancial operating expenses

MEF receives such data continuously during the year, enters them into their computer system, and produces routine periodic reports. At the end of each year, MEF prepares a comprehensive annual report containing over 50 statistical tables most of which are multipage. The process of data collection and data verification are done centrally in a single data processing facility, which houses the computer and is located in Boston.

MEF started its operation in 1978. Its financial reporting system was then completely batch oriented. Data was input via keypunched cards. The amount of data editing was minimal. In addition, the cross-checking of input data across forms submitted by the grantees could not be done easily due to the batch characteristic of the system. The data analysts wanted to have an on-line query capability, which was totally absent in the batch system. As a result, there was an average time lag of 6 weeks from when the data first entered the system and to when it ultimately became clean. This often led to delays in producing reports.

In early 1984, MEF contacted the system consulting team HPC to convert the batch system into an on-line system. After some preliminary investigations, HPC proposed a new system called Financial Reporting System to address the problems. The system would consist of four major subsystems:

   a. Data Entry/Edit/Display

   b. Data Validation

   c. Database Administration

   d. Annual Report Generation

Subsystem (a) will enable the user to call up a blank form for entering new data or an already entered form for editing the data. After each session, the form will be stored in a temporary data store. This subsystem will interface with subsystem (b) via the form number and the grantee ID. This latter subsystem will perform extensive data validity checks to ensure the accuracy of data. It will match appropriate data values both within a given form and across related forms. Subsystem (c) will store a validated form in a relational database. The user will be able to retrieve necessary data from the database and generate ad hoc reports in response to queries. Subsystem (d) is a batch process and will be run only once a year after all the forms data are entered in clean form into the database.

The three feasibility issues for the system are easily satisfied. In fact, the data analysts of MEF especially liked the idea of being able to enter and edit data interactively.

4.5 KEY WORDS

The following key words are used in this chapter:

batch processing	operational scenario
database	order processing
data flow diagram	sales management
feasibility study	subsystem
inventory control	transaction processing

REVIEW QUESTIONS

1. What are the main findings of HPC regarding the problems at Toy World?

2. Suppose that you are an employee of HPC and you want to interview the MIS manager at Toy World to gather information about the data processing operations in general and the order processing system in particular. Write a memo addressed to the MIS manager in order to set up the meeting. Give an outline of the issues you want to discuss at the meeting.

3. Describe the three major subsystems and their interfaces with one another for the system proposed for Toy World.

4. Do you think the rapid growth of Toy World is partially responsible for its current problem?

5. What specific financial constraints are imposed on the system study by Toy World?

6. What problems did MEF face because of the batch nature of its financial reporting system? Why can an interactive on-line system address those problems?

7. What is an operational scenario? What purpose does it serve in describing a problem?

8. The system consultant HPC has provided operational scenarios for their client MEF but not for their other client Toy World. Explain this difference.

9. In the Financial Reporting System, why is a form validated completely before it is entered into the forms database?

10. How can a user check the status of a form stored in the database?

11. Why does FRS provide two levels of user access, namely, USER and ADMIN?

12. What is the difference between data edit and data validation in FRS?

13. Describe FRS's subsystems and their interfaces.

14. Can HPC meet the economic feasibility criteria imposed by MEF? Explain your answer.

5

System Analysis: Theory

5.1 INTRODUCTION

System analysis carries the task started under problem definition and feasibility study further (see Chapter 3). Thus, the difference between these two phases is one of degree only, not of kind. The system analysis starts with the level 1 data flow diagram and the objectives of the proposed system, both of which are now available from the previous phase of problem definition and feasibility. The end product of the system analysis phase is a complete layout of the logical system. During this phase the system team develops a series of more detailed data flow diagrams, a data dictionary, and a brief cost/benefit analysis. Upon approval from the customer, the system team will convert the logical system into the physical system; this conversion occurs during the design phase.

System analysis uses the same fact gathering techniques as does the first phase of problem definition and feasibility. Different levels of personnel in the customer's company are interviewed at a more in-depth level. Normally, the interview proceeds from high-level employees to low-level employees. The former can provide only an overview of the operations of the company, whereas the latter can supply lengthy details on individual parts. The system team needs all these details in order to refine the data flow diagrams and to prepare the data dictionary. For that same reason, the team should also review all the available operations manuals and reports related to the proposed system. Step by step, the logical system is more fully defined.

5.2 DEFICIENCIES OF THE CURRENT SYSTEM

The system team must study existing system carefully for two reasons:

a. To identify its deficiencies and bottlenecks, if any, in the flow of its information

b. To determine if anything can be salvaged from the current system

The first item helps the team avoid pitfalls of the present system; the second item offers some simplifications during the design and implementation phases. Since the current system is an existing physical system as opposed to a conceptual logical system, its system flowcharts and operating procedures are the vehicles to provide information on the system. The system team analyzes these documents to determine the data collection, data processing and storing, and report generation, as practiced within the existing system.

We are assuming that the current system is well documented so that its system flowcharts and operating procedures already exist. However, this may not be the case. In fact, lack of adequate documentation of the existing system may be the source of the present problem. In such cases, the system team must prepare the necessary system flowcharts and some basic operating procedures on the basis of the information gathered during the in-depth interviews and should have these checked by the interviewees for accuracy. A system flowchart is a graphical representation of a physical system and is the most useful form of documentation of the system. It must be the starting point for studying the existing system. The operating procedures provide the supporting narrative for the system flowchart.

Multiple levels of system flowcharts are necessary to study the existing system. Ideally, there should be one high-level flowchart describing the entire system and at least one detailed flowchart for each subsystem. As an example, Figure 5-1 represents the high-level system flowchart of an order processing system for a wholesale distributor. It provides a basic overview of entire system. The system consists of two major subsystems—order checking and inventory control. Figures 5-2 and 5-3 provide detailed flowcharts for these two subsystems. The order checking subsystem (see Fig. 5-2) uses the inventory master file and the customer master file in the processing of orders. But Figure 5-2 does not contain any details to control inventory. Hence, Figure 5-3 is needed to show how the inventory master file is maintained. We assume that the customer master file is maintained by another system (e.g., Accounts Receivable) outside the order processing system. Hence no flowchart is shown to describe the maintenance of that file.

The system flowchart indicates the various reports, files, and transaction documents used in an existing system. (If the system is manual, the files may be in filing cabinets.) Collecting and analyzing copies of the various reports, file contents, and transaction documents in the context of the systems

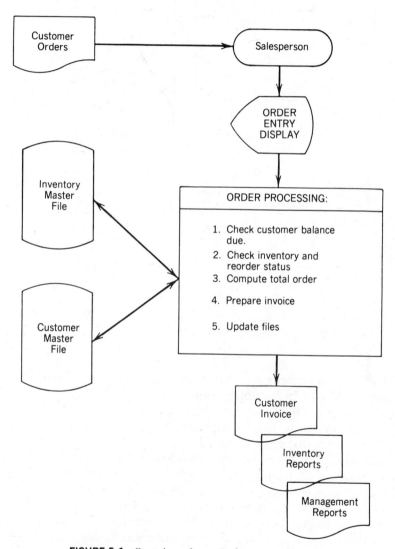

FIGURE 5-1 Overview of an order processing system.

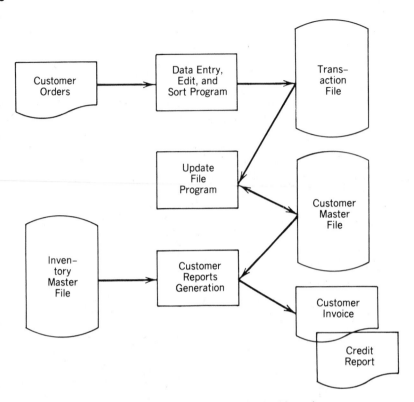

FIGURE 5-2   Detailed flowchart of order checking subsystem.

flowchart leads to a thorough understanding of the actual processing involved
in transforming data into usable information.

Once the system flowcharts and the operating procedures are reviewed,
the system team identifies the discrepancies between the information required
to run the system efficiently and the information available from the existing
documentation. This enables the team to identify deficiencies in the existing
system. For instance, the team may find that certain documents are produced
in multiple copies unnecessarily or that the signature of the president of the
company is required on a large number of documents, which creates a bottleneck in the current flow of information. Interviews with managers may
indicate that data critical for making important decisions are either not collected or are spread over several reports making it difficult to retrieve them
in a timely manner. Some of the operating procedures may be found to be
needlessly repetitive and therefore slowing down the system. Such bottlenecks and deficiencies should be listed, documented, and, if possible, identified on the system flowcharts. These deficiencies should include both *what*
information, technology, and personnel are included or lacking in the system
and *how* the information, technology, and personnel are organized and in-

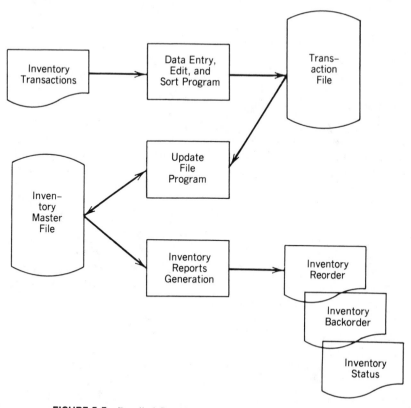

FIGURE 5-3   Detailed flowchart of inventory control subsystem.

terrelated throughout the system. Figure 5-4 gives a sample of definitions and examples of such deficiencies; see Wetherbe ([3], Chapter 5) for more examples.

5.3   EXPLOSION OF DATA FLOW DIAGRAMS

Fortified with the knowledge of problems with the existing system, the system team concentrates on refining and detailing the proposed system. The high-level data flow diagram prepared during the first phase is the starting point here. This diagram consists of five distinct entities:

☐ Data flows
☐ Data stores
☐ Processes
☐ Source(s)
☐ Destination(s)

What Is Lacking?	How Is It Lacking?
Information: Needed information is not collected, and/or unnecessary information is taken; e.g., an inventory control system does not have an optimal reorder policy.	*Information:* Storage and retrieval of needed information are inconvenient; e.g., users have to depend on data processing to generate inventory status report.
Technology: Improper system functions are implemented; e.g., data entry and edit are done via keypunched cards.	*Technology:* The hardware and software are not organized to process the data in the most efficient manner; e.g., applications still use programs written in unstructured COBOL.
Personnel: Appropriate personnel are not available; e.g., well-trained word processing operators are missing.	*Personnel:* Improper recruitment and/or supervision still persist in the company; e.g., job classifications are poorly prepared.

FIGURE 5-4   Deficiencies of an information system.

Conceptually, each process represents a functional capability to be implemented by the system. A process at the highest level can be further detailed and broken down into multiple processes at the next lower level. This type of decomposition can be repeated until the system team feels that enough details have been provided so that any further subdivision of a process to the next lower level will involve a physical implementation of the process.

When each process is broken down in this manner, the other entities in the data flow diagram are affected. Especially, the data flows and the data stores increase in number. Some new sources and destinations may also arise. As a result, the high-level data flow diagram, which is called the level 1 data flow diagram, produces two or more level 2 data flow diagrams. This procedure is called *explosion* of data flow diagrams. Each process at level 1 is exploded into two or more processes at level 2 and produces a level 2 diagram for that process. The processes at level 1 are numbered 1, 2, 3, ... Process 1 is then exploded into processes 1.1, 1.2, 1.3, . . . at level 2. Process 1.1 is then exploded into processes 1.1.1, 1.1.2, 1.1.3, . . . at level 3, and so on. This numbering scheme is consistently followed throughout the explosion process in order to indicate the hierarchy of decomposition of each process. The number of a process at any level determines its whole line of "ancestors."

The procedure of exploding a data flow diagram increases the number of processes in the entire system almost exponentially. This increases the complexity of the diagram since the data flows and the data stores also tend to increase in number due to the explosion. So, the obvious question is, How many processes can be included in a data flow diagram at any level? Various studies done on human cognitive levels suggest that 7 ± 2 processes should be the range for a data flow diagram at any level. Thus, a diagram should contain five to nine processes. If a diagram needs more than nine processes, it should be broken down into multiple diagrams. Likewise, if a diagram has four or fewer processes, perhaps it should be made a part of another diagram at the same level.

The principle behind the structured analysis is to break a system into its major subsystems. Each process at level 1 is normally taken to represent a subsystem. Thus, each level 2 data flow diagram can be regarded as the level 1 diagram for that subsystem. If explosion continues beyond level 2, these lower level diagrams are regarded as more detailed conceptual representations of the various subsystems. However, data flow diagrams are rarely exploded beyond level 4. Since a process provides the fundamental ingredient of a subsystem, it is taken as the basis for explosion.

Figure 3-2 of Chapter 3 is a level 1 data flow diagram for a sample system. We now illustrate the process of explosion of a data flow diagram by providing two data flow diagrams at level 2, which explode processes 1 and 2 from level 1, and two data flow diagrams at level 3, which explode processes 1.1 and 1.2 from level 2. Figures 5-5 to 5-8 are self-explanatory. We note that during the explosions four new data stores are created:

```
EDIT CHECKS
ENTERED DATA
VALID USERS
FORM TEMPLATES.
```

They were not shown in Figure 3-2 earlier because processes 1 and 2 were not sufficiently detailed at that stage. Additional data flows also appear in the exploded versions.

5.4 ALTERNATIVE SOLUTIONS AND RECOMMENDATIONS

The layout of the complete logical system has now been done. The next step is to determine alternative ways of implementing the system. This leads us to finding alternative solutions for the problem. It is customary in system development to provide two or three alternative ways to implement the logical system. The system team also describes the advantages and disadvantages of each alternative and concludes with its own recommendations as to which alternative should be selected. Of course, the customer has the final say in determining which alternative to choose.

The system team should consider three possible alternatives

a. Completely or predominantly manual system
b. Partially batch and partially on-line system
c. Completely on-line system

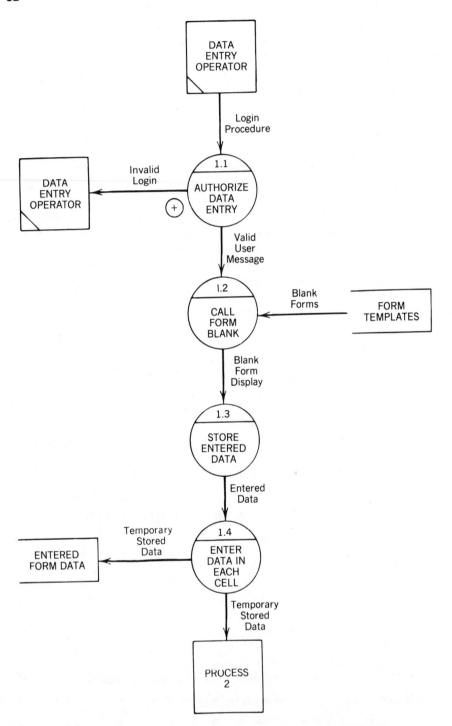

FIGURE 5-5 Explosion of process 1 – ACCEPT ENTERED DATA.

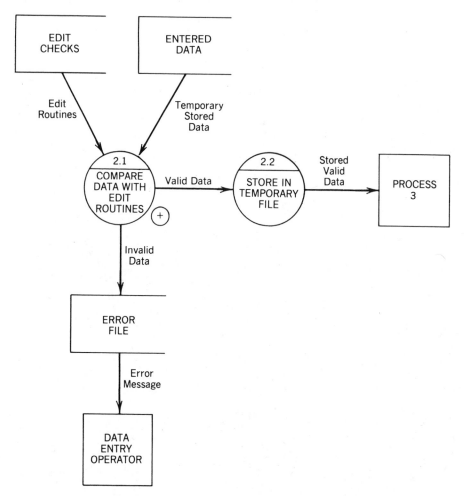

FIGURE 5-6 Explosion of process 2 – VERIFY ACCEPTED DATA.

From an implementation standpoint, these options are progressively expensive and complex to design. But from the standpoint of functional capability, these options are progressively versatile.

The level 1 data flow diagram is used to generate alternative ways of implementation. The key factor is to look at each process, examine its time sensitivity, and then combine the processes into separate groups according to their time sensitivity. For example, processes that need a response time of 1 minute or less may be classified as on-line, those that have a response time of up to a day may be regarded as batch, and so on. The systematic way to address the issue of grouping individual processes is by using the concept of automation boundary. We shall discuss this topic in Section 5.5.

One final comment is in order. When we determine the alternative so-

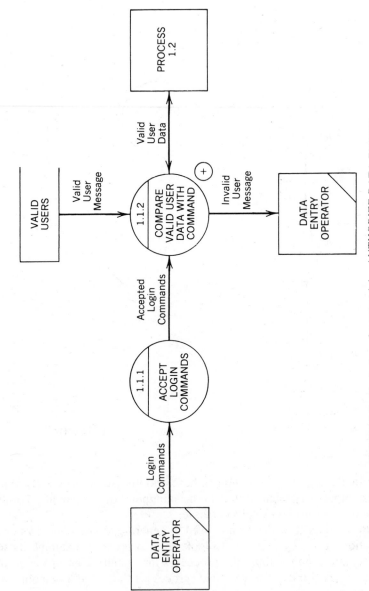

FIGURE 5-7 Explosion of process 1.1 – AUTHORIZE DATA ENTRY.

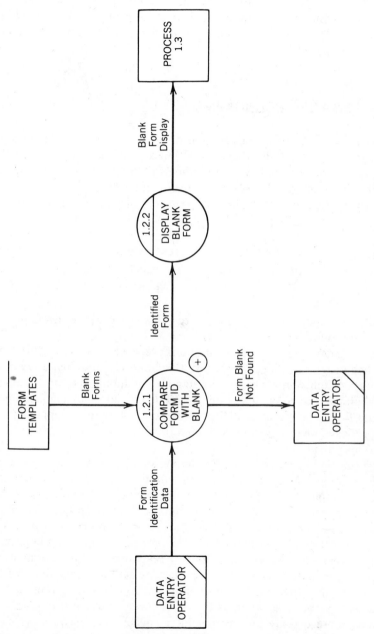

FIGURE 5-8 Explosion of process 1.2 – CALL APPROPRIATE FORM TEMPLATE.

lutions, we indeed look into the physical system rather than the logical system, since a solution invariably refers to the implementation process. Thus, we have to take a glimpse into the physical system while we are still at the analysis stage. But this interruption is justified, because otherwise the user cannot select an alternative solution on the basis of which the subsequent design phases will start.

5.5 AUTOMATION BOUNDARIES

An *automation boundary* is a partition or grouping of processes in a data flow diagram such that all processes within a given boundary, that is, belonging to the same group, have similar time sensitivity. Processes are divided into different categories based on the user requested response times. One such possible breakdown into three different response time categories is given below:

a. Immediate response category, i.e., response time less than 10 seconds
b. Intermediate response category, i.e., response time of an hour or less
c. Overnight response category, i.e., response time of at most 24 hours

It should be noted, however, that the response time categories depend on the system requirements.

From the standpoint of implementation, processes in category (a) should use on-line programs, those in category (b) should use terminal driven modules with a batch option, those in category (c) should use the traditional batch programs, and so on. Thus, imposing one possible automation boundary on a data flow diagram at level 1 provides one alternative implementation of the proposed logical system. The data flow diagram works here as a significant design aid.

Using the timing requirements of the various processes as a guide, it is possible to impose a number of different automation boundaries on a given data flow diagram. Each automation boundary suggests a possible physical system. Response time is the sole criterion for including processes within a given automation boundary. When including multiple processes within a single automation boundary, the response time for that boundary is determined by the member process with the shortest response time requirement. For instance, if a boundary includes four processes with response times of 3, 5, 6, and 10 seconds, then the boundary must satisfy a response time of 3 seconds. This is so because whereas the other three slower processes may be performed more quickly than is absolutely necessary, the process with a response time of 3 seconds cannot be forced to wait any longer for the slower ones. Thus, an automation boundary is designed to satisfy the most demanding response time requirement within its borders, making the other less demanding processes possibly more active than they need be.

Let us now prepare two alternative solutions of the system described in Section 3.8. We impose two separate automation boundaries on Figure 3-2 as follows:

a. Processes 1 and 2 have a response time of 5 seconds or less and hence are performed on-line. Process 3 has a response time of at most 24 hours, whereas process 4 has a response time of a day or more. Consequently, processes 3 and 4 can be implemented via batch programs.

b. Processes 1, 2, and 3 have response time requirements of 5 seconds, 5 seconds, and 30 seconds, respectively, because the user wants to have the master file updated continuously. Process 4, however, has a response time of a day or more. Hence processes 1, 2, and 3 are implemented on-line with a response time of 5 seconds, whereas process 4 is still run in batch mode.

Figures 5-9 and 5-10 show these two automation boundaries, respectively.

5.6 DATA DICTIONARY

A *data dictionary* is a repository of data about data. During the system analysis phase the level 1 data flow diagram is exploded into a number of more detailed-level data flow diagrams. Let us call the latter micro data flow diagrams. A micro diagram has three principal entities:

a. Data flows
b. Data stores
c. Processes

Processes operate on the data flows and are determined by the system team. Data stores are merely collections of data flows related to the same category of items. However, data flows originate inside the user organization and are thus external entities. Very often different people within the organization talk about the same data flow using different terms. It is necessary to clarify these ambiguities at the initial stage to avoid future complications. For example, at the programming stage, it is necessary to refer to a given data element by a unique name. The data dictionary is the right vehicle to accomplish that task.

The data dictionary is prepared as follows:

1. Make a master list of all the data flows, data stores, and processes appearing in the micro data flow diagrams.
2. For each data flow, prepare an entry using the format of Figure 5-11.
3. For each data store prepare an entry using the format of Figure 5-12.
4. For each process prepare an entry using the format of Figure 5-13.

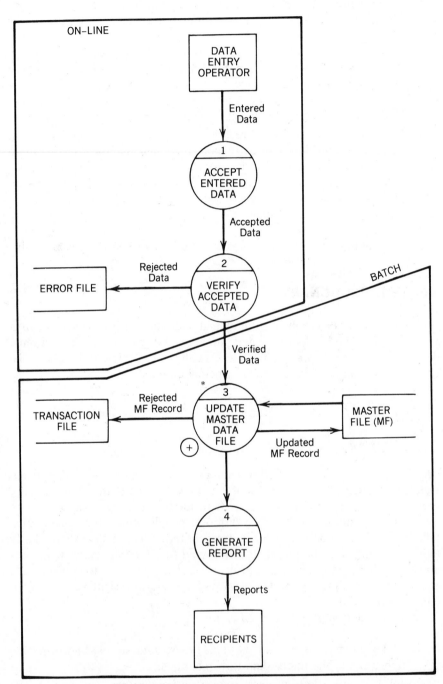

FIGURE 5-9   Automation boundary 1 for Figure 3-2.

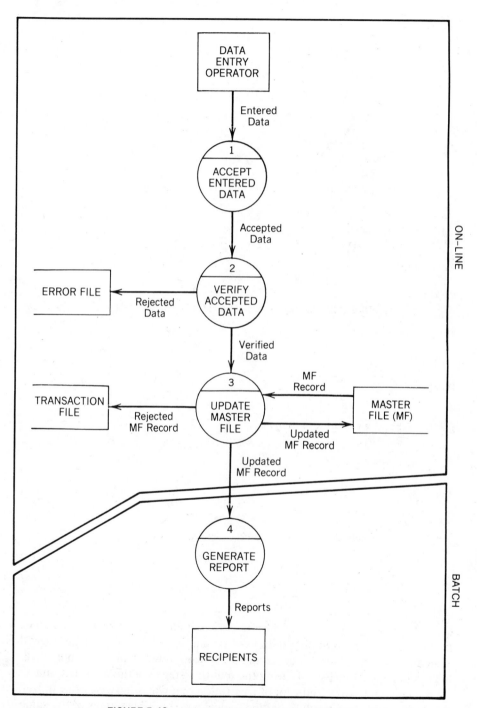

FIGURE 5-10 Automation boundary 2 for Figure 3-2.

Name:

Aliases:

Description:

Format:

Location:

FIGURE 5-11   Format of data flow entry in data dictionary.

Name:

Aliases:

Description:

FIGURE 5-12   Format of data store entry in data dictionary.

Name:

Number:

Description of process logic:

Input:

Output:

FIGURE 5-13   Format of process entry in data dictionary.

The "aliases" in Figures 5-11 and 5-12 include all the alternative names by which the corresponding data flow or data store is called in the logical system. Since the system is developed by a team rather than by an individual, aliases indeed arise. By capturing them at this stage via the data dictionary, the system team can avoid a lot of frustration that will otherwise arise during the design stage as a result of having to deal with multiple names of the same data element. Also, during the design stage, each data flow entry in the data dictionary has to be expanded into the data element level; similarly, the data stores must be converted into data files. The entries in the data dictionary prepared during the analysis phase provide valuable help at that time. In fact, the "location" item in Figure 5-11 directly indicates which data flow belongs to which data store. If multiple data stores appear in the "location" cell, it indicates a potential problem of updating multiple files when the corresponding data flow changes.

A number of data dictionary software packages is commercially available. Some are associated with a specific data base management system. Others are more general; many offer optional links to a variety of data base management systems. Some firms have even written their own customized data dictionary software. If such software is unavailable, a manual system should be used to simulate the software.

Many advantages associated with a data dictionary are derived from the ability to process information about each data flow, data store, and process separately. Simply listing all the data elements on a sheet of paper will not do, since it is difficult to deal with the individual elements when they are presented in list form. To maintain a semblance of direct access, information related to each data element should be recorded on a separate 3″ × 5″ filing card. The stack of these completed cards then become the data dictionary.

We close this section with a brief description of a software called Visible Analyst Workbench, which automates the preparation of data flow diagrams and data dictionary. This is marketed by Visible Systems of Newton, Massachusetts. Version 2.2 has been made available in late 1987.

The software consists of three modules, called Tool A, Tool B, and Tool C. Tool A, the Visible Analyst, is the fundamental diagramming tool. It allows the user to draw various graphics such as data flow diagrams, system flowcharts, and program structure charts. The user can generate a variety of new symbols by using the custom symbol generator containing 21 of the most common data flow diagram, system and program flowchart, and presentation symbols. Tool B, the Visible Rules, guides the user in designing systems. It interfaces with Tool A and provides balancing among multiple levels of data flow diagrams, checks for consistency, and generates warning and error messages such as "data flow is unnamed," and "data flow is dangling." Tool B checks for the following six items:

a. Unnamed symbols or data flows
b. Unnumbered processes
c. Dangling symbols and lines
d. Data flows that are not attached to a process
e. Data flows that are not carried down to lower levels
f. Unnamed data flow attached to a data store

Tool C, the Visible Dictionary, adds a central repository for information related to the diagrams. It interfaces with Tools A and B. Tool C automatically loads all key information from a diagram such as a DFD. Examples of information automatically loaded are: process name, process number, location of data flow, and type of entry (e.g., data flow, data store, process, and data element). Tool C also allows the user to perform selective and wildcard searches of the data dictionary and generate from it three types of reports: detailed, summary, and cross reference. The dictionary thus serves as a central definition catalog for all the diagrams.

The user can access the entries in the data dictionary (Tool C) from the graphics screens (Tool A). This feature provides the user greater flexibility to edit entries while the processes and data flows are being created.

5.7   DEVELOPMENT AND MAINTENANCE COSTS

In Section 3.9 we discussed the three feasibility issues, one of which was economic feasibility. Economic feasibility involves a determination of whether the system development project can be implemented within the budget allocated by the customer for that purpose. This requires an estimate of the total cost for the project. Each of the five phases of the system life cycle incurs a cost. There are two types of cost:

   a. Development cost
   b. Maintenance cost

5.7.1   Developmental Cost

Developmental cost is a one-time cost. It starts with the inception of the project and continues through the implementation phase. Once the system is operational, the developmental cost ceases.

Developmental cost is broken down into the following four major components:

   1. Hardware. Hardware consists of the computer equipment and all peripherals. Thus, it includes the CPU and memory, disk drives, tape drives, printers, and terminals.

   2. Software. Software consists of all programs used in the operation of the system. It includes the operating system for the computer, all application programs written by the programmers, and any additional software that is purchased from vendors such as database management systems and special packages like forms generator and menu-driven screen generators.

3. **Data Communications.** Data communication consists of both hardware and software necessary for transmitting data from one location to another. Since not all systems require a data communication segment, it is appropriate to consider it as a separate item. Besides the computer hardware already discussed above, the data communication hardware includes equipment such as modems, communication channels with appropriate baud rate, communication controllers, multiplexers, and concentrators. Data communication software consists of programs that make the sending and receiving of data via communication lines possible in an error-free manner. Mostly, various vendors provide different communication software packages. See Lucas ([2], Chapter 10) for an overview of data communication issues.

4. **Personnel.** Personnel includes all the team members and any other supervisory personnel. If any consultants are involved, the consulting fees should be included. Clerical people used for temporary help such as data entry, generation of graphics (e.g., data flow diagrams, system flowcharts, structure charts), typing reports, and so on must also be included.

Davis ([1], pp. 321–323) comments as follows on how to estimate the total developmental cost:

> It is difficult to estimate the cost of a system. It is much easier to estimate the cost of each of the system components, and then sum these results; the total cost of the system is, after all, nothing more than the sum of the costs of its parts. Thus, the first step in developing a cost estimate is to break the problem into as many pieces as possible. During the feasibility study, this subdivision might be limited to the steps in the system life cycle, yielding an estimate that is perhaps accurate to ±50 percent. Later, the personnel, equipment, and material costs associated with each step might be estimated separately. During detailed design, those personnel costs might be subdivided into programmers, operators, clerks, and others. Eventually, the analyst may be able to break the work of the programmers down to individual modules on a hierarchy chart and generate a more accurate cost estimate; the smaller the module, the easier it is to estimate the cost.

> On most systems, the bulk of the cost is concentrated in a few of the system components. By evaluating these few key elements carefully, it is possible to develop a very accurate cost estimate with a minimum of effort. If the handful of factors that account for 80 or even 90 percent of a system's cost can be identified and accurately estimated, ballpark figures or even guesses may be perfectly adequate for the other factors.

5.7.2 Maintenance Cost

Maintenance cost includes all expenses needed to continue the operation of the system. It is thus an ongoing or recurring cost: It continues throughout the life of the system. Maintenance cost can be broken down into the following components:

1. **Hardware.** Hardware cost includes the usage and depreciation of the equipment, rental or lease charges if appropriate, and the maintenance contract for scheduled and on-demand maintenance. It also includes any enhancement needed for continued smooth operation of the system. Additional memory or extra secondary storage space also belongs to this category.

2. **Supplies.** Supplies include forms, paper, keypunch cards, tapes, disk packs, and so on. This is a fairly small portion of the total maintenance cost.

3. **Personnel.** Personnel includes computer operators who run the system, programmers who maintain and modify the system, data entry operators, clerks who supply source documents and distribute final reports, and management responsible for running the system. Any overhead cost should also be included here.

Normally, the maintenance cost is only a small fraction of the developmental cost. Consequently, we often say that the developmental cost is an investment, whereas the maintenance cost is an expense.

5.8   TANGIBLE AND INTANGIBLE BENEFITS

When a company makes an investment, it always looks for the return from that investment. If it finds that the amount of capital outlay cannot be recovered in a few years, it would not venture into the investment. The same situation holds in case of allocating a budget for developing a system. The customer wants a cost justification for the developmental cost. It is the responsibility of the system team to provide this cost justification. This leads us to the issue of benefits from developing a system.

The developmental cost, and perhaps the maintenance cost as well, are justified by enumerating and analyzing the benefits rising out of the operational system. There are two kinds of benefits:

a. Tangible benefits
b. Intangible benefits

5.8.1   Tangible Benefits

Tangible benefits are those to which a dollar value can be assigned. For example, if a new sales information system leads to an increase in sales volume, the additional revenue coming out of sales is a tangible benefit.

Similarly, if an improved inventory control system results in more efficient handling of the items in the inventory, then the company may be able to free up some cash that used to be tied up in unused inventory. By investing this extra cash, the company will be able to generate additional investment income, which can be designated as a tangible benefit of the inventory control system.

5.8.2   Intangible Benefits

Intangible benefits are those that cannot be measured in hard dollar values. For example, suppose that a retailing company is getting numerous complaints from its irate customers who claim that amounts already paid are still shown on their monthly bills as unpaid. This leads to a loss of customers and also a bad public image of the company. If the company decides to implement an improved customer invoice system, the problem can be remedied. However, it is difficult to assign a hard dollar value to the benefits. For example, how can you estimate the amount of extra revenue that would come from a new customer? Or, how would you assign a dollar value to an improved public image or to goodwill? Obviously, the system team can only make estimates in such cases. But there is a risk that these estimates may be wrong. Often an intangible benefit is associated with a soft dollar value.

5.9   COST/BENEFIT ANALYSIS

The assessment of economic feasibility (see Section 3.9) of a system development project raises the fundamental question; Can the benefits outweigh the costs? Regarding the development cost as a capital investment and the maintenance cost as an operating expense, the customer regards the budget allocated for the project as an investment risk. The benefits accruing from the project represent the return. Accordingly, the customer expects that the return should outweigh or at least justify the risk. However, it does not always happen that way. There are projects on which the costs cannot be justified in hard dollar value of the tangible benefits. The purpose of the cost/benefit analysis is to examine this type of trade-off situation in a systematic and quantitative manner and then determine if the tangible and intangible benefits together can justify the project costs.

The process of cost/benefit analysis consists of the following steps:

a. Estimate the total cost as the sum of the developmental cost and the maintenance cost. The developmental cost, at this early stage, is only an estimate and can go up to 50 percent more. The maintenance cost can be estimated as 10 percent of the developmental cost if no better figures are available.

b. Estimate the total benefit as the sum of tangible and intangible benefits. The tangible benefits are easier to compute and can be translated into hard dollars. The intangible benefits are more difficult to compute since they are not normally expressible in terms of hard dollars. Use some rationale to assign a dollar value to these benefits. For example, suppose that a new customer invoice system is expected to bring in new customers. How can we assign a dollar value to the business brought by a new customer? Examining historical records, we can estimate the dollar value of "average" customers for the last 5 years, say. Then we can use this number as our estimate. However, any such rationale must be described and documented in the cost/benefit analysis report.

c. Estimate the life span of the system. Usually, a 5-year span is taken as standard.

d. Using some appropriate and accepted accounting method, perform a present value analysis of the total benefit to determine the payback period. The *payback period* represents the total number of whole and fractional years after which the accumulated benefits will exceed the initial investment made in the form of developmental cost.

We now outline a methodology to address step (d). Since any amount of money not sitting idle generates income (e.g., in the form of interest), money is said to have a time value. Thus, an amount of $1,000 today will be worth more a year later if it is invested at a certain rate of interest. Looking at the situation in the reverse order, we find that in order to get a return of $1,000 1 year later, we should invest less than $1,000 now. Now, regarding the return as the total benefit and the investment as the developmental cost, we can say that in order to reap a system benefit of $1,000 1 year later, the system development cost must be less than $1,000. But how much less? The answer depends on the rate of interest. Let us now formulate a quantitative method to handle this situation. We shall distinguish between two cases according to whether the annual ongoing maintenance costs are not considered or are considered.

Suppose that we invest an amount P at an annual compound interest rate of i for n years. Then the accumulated amount F at the end of n years is given by the formula

$$F = P(1 + i)^n$$

$$\therefore P = \frac{F}{(1 + i)^n}$$

Hence for given values of i and n, we can compute P when F is known. In our methodology given below, F represents the annual benefit and P represents the present value of that benefit for each value of n ranging over the total life span of the system.

Case 1: Ignore the ongoing annual maintenance cost

a. Let the system life span be t years, where t is a whole number.

b. Let $B_1, B_2, \ldots, B_t$ be the annual benefit of the system at the end of the first, second, $\ldots, t^{th}$ year, respectively.

c. Let

$$P_j = \frac{B_j}{(1 + i)^j}, \qquad j = 1, 2, \ldots, t$$

Then P_j is the present value of the benefit B_j available in year j, where $j = 1, 2, \ldots, t$.

d. Find the number N, which can be a whole number plus a fraction, such that the cumulative value

$$P_1 + P_2 + \cdots + P_N$$

is approximately equal to the developmental cost C. Then the payback period is N.

In general, N is less than t. Thus, the customer expects to realize a tangible benefit during the life span of the system. However, there are situations in which the customer agrees to develop a system knowing that the payback period will exceed the total system life.

Case 2: Include the ongoing annual maintenance cost

a. Let $M_1, M_2, \ldots, M_t$ be the annual maintenance cost for the first, second, $\ldots, t^{th}$ year, respectively.

b. Retain assumptions (a) and (b) under case 1.

c. Since for year j, an amount M_j must be spent for maintenance, the *net* benefit for that year is

$$B_j - M_j, \qquad j = 1, 2, \ldots, t$$

d. Repeat steps (c) and (d) under case 1 by replacing B_j with $B_j - M_j$, $j = 1, 2, \ldots, t$. Then, N is the payback period.

We are assuming that

$$B_j > M_j, \qquad j = 1, 2, \ldots, t$$

since otherwise the system costs will never payback.

We now illustrate the method with an example.

Example. Let us suppose that a company intends to spend $10,000 to implement an improved customer invoice system. The system team estimates that the total annual benefit for the next 6 years will be, respectively, $6,000, $5,500, $5,500, $3,000, $3,000, and $2,000. A significant part of the benefit will be of an intangible type arising out of new customers due to improved public image of the company. However, it is expected that the customer base will stabilize by the end of the third year. Hence, there will be a sudden drop in the total benefits starting with the fourth year. Consequently, the total benefits during the last 3 years are almost half of those during the first 3 years. We now compute the payback period.

Case 1: Ignore ongoing maintenance cost

Here

$t = 6$

$B_1 = \$6,000$

$B_2 = \$5,500$

$B_3 = \$5,500$

$B_4 = \$3,000$

$B_5 = \$3,000$

$B_6 = \$2,000$

Assume $i = 14$ percent.

Figure 5-14 shows the cost/benefit analysis in this case.
We notice that between the second and the third year the company realizes a total benefit exceeding the initial investment of $10,000. During the third year an additional amount of $506.07 (10,000 − 9,493.93) is required to make up the initial investment. However, the total present value of the third year's benefit is

$$P_3 = 3,716.22$$

Year (j)	Benefit (B_j)	Interest (1 + i)^j	Present Value (P_j)	Cumulative Present Value
1	$6,000	1.14	$5,263.16	$ 5,263.16
2	5,500	1.30	4,230.77	9,493.93
3	5,500	1.48	3,716.22	13,210.15
4	3,000	1.69	1,775.15	14,985.30
5	3,000	1.93	1,554.40	16,539.70
6	2,000	2.19	913.24	17,452.94

FIGURE 5-14   Example of cost benefit analysis: Case 1.

Year (j)	Benefit $(B_j - M_j)$	Interest $(1 + i)^j$	Present Value (P_j)	Cumulative Present Value
1	$5,000	1.14	$4,385.96	$ 4,385.96
2	4,500	1.30	3,461.54	7,847.50
3	4,500	1.48	3,040.54	10,888.05
4	2,000	1.69	1,183.43	12,071.47
5	2,000	1.93	1,036.27	13,107.74
6	1,000	2.19	456.62	13,564.36

FIGURE 5-15 Example of cost benefit analysis: Case 2.

Using linear extrapolation, we find that $506.07 is realized in the fractional part 0.14 of the third year since

$$\frac{506.07}{3716.22} = 0.14$$

Thus, the total payback period N is 2.14 years.

Case 2: Include ongoing maintenance cost

In addition to all the assumptions in case 1, we assume that the annual maintenance cost is $1,000. Hence we take

$$M_j = 1000, \quad j = 1, 2, \ldots, 6$$

Figure 5-15 gives the cost/benefit analysis in this case. Proceeding as in case 1 above, we find that the payback period N is 2.71 years.

5.10 CONCLUSIONS

As a result of phases 1 and 2, we get an in-depth understanding of the requirements of the proposed system. The end product is a report describing the following:

a. Problems, deficiencies, and bottlenecks of the present system
b. Objectives of the proposed system
c. Description of the proposed system (only *what* functions will be achieved, not *how* to achieve them)
   ☐ List of subsystems
   ☐ List of interfaces among the subsystems
d. Multilevel data flow diagrams showing
   ☐ Processes
   ☐ Data flows
   ☐ Data stores

e. Data dictionary providing entries for *all* processes, data flows, and data stores

f. Feasibility issues

g. Cost/benefit analysis

h. Alternatives and recommendations

Since we are using both data flow diagrams and system flowcharts, it is worthwhile to note their similarities and differences:

1. Similarities
   a. Both are graphic tools.
   b. Both describe the flow of data through the system and the logic used therein.
   c. Both show the data to be stored.
   d. Both show the major functions to be accomplished by the system.
2. Differences
   a. A data flow diagram is conceptual, and a system flowchart is physical.
   b. A data flow diagram is independent of implementation, and a system flowchart is implementation bound.

Finally, we provide a checklist of items that should appear in the findings report concluding phase 2 of the study. Clearly, some of the items are just carryovers from phase 1 in order to provide continuity. The checklist is given in Figure 5-16.

1. Introduction
   Thank you note for acceptance, recap of problem definition.
2. Scope and Objectives
   List of capabilities to be provided with a brief (one or two sentences) description of each capability.
3. System Flowcharts for Current System
   Identity bottlenecks and deficiencies in the flowchart.
4. List of Deficiencies in the Current System
5. Exploded DFDs at Level 2
   Expand each level 1 into component level 2 diagrams.
6. Data Dictionary
   Include *all* data flows, data stores, and processes.
7. Alternatives and Recommendations
   Provide *at least two* alternatives and recommend one with reasons.
8. Brief Cost/Benefit Analysis
   Indicate payback period.

FIGURE 5-16   Checklist for findings report of phase 2.

5.11   SUMMARY

This chapter discussed the second phase of a system development study, namely, system analysis. System analysis examines in more details the functional capabilities of the proposed system and uses specific graphic tools to document the findings. Two primary fact gathering techniques used in this phase are (1) interviewing and (2) reviewing earlier studies and current operating procedures. As a result of this process, the system team determines the deficiencies of the current system, documents them in a list of deficiencies, and identifies them in one or more system flowcharts of the existing system. Since the current system is a physical system rather than a logical one, system flowcharts rather than data flow diagrams are the appropriate graphic tools to describe the system. The purpose of identifying the current system's defects is to ensure that they are not repeated in the proposed system.

The next task in this phase is to further describe the proposed system by preparing more data flow diagrams. This is called the *explosion* of data flow diagrams, whereby each process in the level 1 diagram is detailed into one or more data flow diagrams at level 2. If necessary, a process in a level 2 diagram can be exploded into one or more level 3 data flow diagrams, and so on. Each distinct data flow, data store, and process in the detailed data flow diagrams is separately documented in a data dictionary. The entries for data flows and data stores are later used in the design phases to prepare the fields and records format for the data files in the proposed system. The entries for the processes in the data dictionary are used to prepare the logic for programs during the design and implementation stages. The data dictionary also helps in identifying multiple locations (i.e., files) for the same data element. It thus cautions the system team toward the problem of multiple file updates to maintain data integrity.

The system team examines the level 1 data flow diagram and groups different processes by their respective response time requirements. For example, the processes can be divided into three response time categories:

a. Immediate response time, say up to 10 seconds
b. Intermediate response time, say up to 1 hour
c. Overnight response time, say up to 24 hours

Based on such categorization, the system team can impose different alternative automation boundaries on the level 1 data flow diagram. Each response time category falls within a single automation boundary. This helps the team to propose several alternative solutions for the problem and also to offer recommendations as to which solution should be selected. Usually three alternatives are provided, one of which is recommended.

The final task during the system analysis phase is to prepare a cost/benefit analysis for the proposed system. The system incurs two types of cost:

a. One-time developmental cost

b. Ongoing maintenance cost

The customer regards the former as a one-time investment and the latter as an operating expense. Also, the latter is only a small fraction of the former. In order to justify the developmental cost as an investment, the system lists the benefits of the proposed system. Here again, there are two types of benefits:

a. Tangible benefits

b. Intangible benefits

A hard dollar value can be associated with the former; the latter usually involves things such as better public relations and improved employee morale.

As part of examining the economic feasibility of the system, the system team raises the fundamental question, Can the benefits outweigh the cost? The cost/benefit analysis tries to answer that question in a quantitative manner. It uses the method of discounted present value of future returns to determine the payback period of the initial investment needed as the developmental cost. At the end of the payback period, the accumulated benefits exceed the developmental and maintenance costs, thereby yielding a positive return on the investment. A detailed example is given here to illustrate the cost/benefit analysis calculations.

The chapter closed with an outline of the topics that are included in the end product, which is a report summarizing the findings from the system analysis phase.

5.12   KEY WORDS

The following key words are used in this chapter:

alternative solutions	operating procedure
automation boundary	payback period
cost/benefit analysis	present value
data dictionary	problem definition and feasibility
data flow diagram	recommended solution
developmental cost	response time categories
explosion of data flow diagram	system deficiencies
intangible benefits	system flowchart
maintenance costs	tangible benefits

REFERENCES

1. William Davis, *System Analysis and Design: A Structured Approach*, Addison-Wesley, Reading, MA, 1983.
2. Henry C. Lucas, Jr., *Analysis, Design, and Implementation of Information Systems*, McGraw-Hill, NY, 1985.
3. James C. Wetherbe, *System Analysis for Computer-Based Information Systems*, West Publishing Company, St. Paul, MN, 1979.

REVIEW QUESTIONS

1. How does the system analysis phase differ from the problem definition and feasibility phase?
2. Why is it necessary to know the deficiencies of the existing system?
3. Why should you use the system flowchart instead of the data flow diagram to identify the bottlenecks of the existing system?
4. What is meant by explosion of a data flow diagram? Why is it needed?
5. Define automation boundary. How can you identify the boundaries in a data flow diagram?
6. What role do the automation boundaries play in providing alternative solutions?
7. Why is it necessary for the system team to provide a definite recommendation for one of the alternative solutions?
8. Define a data dictionary. What is the use of providing aliases in the entry for a data flow?
9. How can you use the data dictionary to identify multiple occurrences of a data flow in the data stores?
10. Explain the two types of cost associated with a system. Why is the developmental cost regarded as an investment?
11. Explain the two types of benefits associated with a system.
12. What does a cost/benefit analysis achieve?
13. What is payback period? Why is it so important?
14. How is the economic feasibility of a system related to the study of its cost/benefit analysis?

6

System Analysis: Case Studies

6.1 RECAPITULATION OF PROBLEM DEFINITION FOR TOY WORLD

In Chapter 4 we discussed the order processing system for Toy World, Inc. The system consulting team, HPC Inc., proposed that the new system should address the following major components:

a. Order processing
b. Inventory control
c. Sales management

Figure 6-1 provides a recapitulation of the total functional capabilities to be provided by the new system. In addition to the components listed above, Figure 6-1 includes production operations and marketing applications that will be impacted by the new system: It outlines the scope and objectives of the new system as proposed by HPC.

6.2 DEFICIENCIES OF THE CURRENT ORDER PROCESSING SYSTEM

After having outlined the functional capabilities of the new system, HPC next analyzes the current system for Toy World in order to pinpoint the system deficiencies and bottlenecks as they exist now. There is minimal

104

Inventory Control

1. To provide fast, accurate service in filling customer orders—avoiding stock outs

2. To minimize the amount of money invested in inventory

Order Entry/Sales Transaction

1. To provide an efficient method of screening and recording customers orders and sales transactions

2. To provide the inventory control system with information on accepted orders so that they can be filled quickly and accurately

Production Operations

1. Production control
2. Production planning
3. Production inventory control
4. Physical distribution management
5. Floor operations
6. Warehouse communications

Marketing Applications

1. Market planning
2. Sales forecasting
3. Sales management
4. Product management
5. Advertising and promotion
6. Market research

Sales Analysis

To provide management with information concerning sales activity and sales trends that are required for effective marketing management.

FIGURE 6-1   Functional descriptions of proposed system for Toy World.

documentation available for the current system. Consequently, based on extensive interviews and actual observation of the information flow in the present system, HPC has prepared a set of detailed system flowcharts. Figures 6-2 through 6-4 represent these flowcharts.

In addition, HPC has made a list of eight deficiencies that exist in the present system:

1. Time consuming hand retyping of invoices (cost per invoice about $10)
2. No finished goods file
3. No control of substitutions (these are made by low-level clerical personnel without proper supervision)
4. Managers out of touch with warehouse control
5. No formal information system (salesperson has no statistics for orders or what is on hand)
6. No backorders capability
7. Decision system minimal (you do not know which toys are selling the most, so you cannot make selection decisions)
8. Manual key punching (risk of errors)

Items 1 through 8 are also shown in the offpage connector symbols on Figures 6-2 through 6-4, which illustrate the deficiencies in graphic form.

6.3 EXPLODED DATA FLOW DIAGRAMS

Figure 4-3 represents the level 1 data flow diagram for the new system. HPC now explodes four of the six processes from level 1 to level 2 diagrams. Processes 3 and 6 are fairly elementary in nature since they merely produce the sales reports and the inventory reports, respectively. Hence HPC does not explode them. Figures 6-5 through 6-8 represent the level 2 data flow diagrams for processes 1, 2, 4, and 5, respectively.

6.4 DATA DICTIONARY

Using Figures 4-3 and 6-5 through 6-8, HPC now prepares the following list of entities used in these five data flow diagrams:

1. Sources

Name	Figure
Customer Service	4–3, 6–7
Management	6–6
Sales	4–3, 6–5

2. Destinations

Name	Figure
Management	4–3
Warehouse	4–3, 6–8

3. Data Flows

Name	Figure
Accepted Trans	6–5, 6–6, 6–7
Change Data	4–3
Change Trans	4–3, 6–7
Customer Change Data	4–3, 6–7
Customer Data	4–3, 6–5
Detailed Invoice	4–3, 6–8
Inventory Change Data	4–3, 6–6, 6–7
Inventory Change Trans	6–6
Inventory Data	4–3, 6–5
Inventory Status	4–3
Inventory Status Report	4–3
Master File Change Data	4–3, 6–7
Master File Data	4–3, 6–5, 6–8
Order Data	4–3
Order Trans	4–3, 6–5
Picking List	4–3, 6–8
Rejected Trans	6–5
Sales Change Data	4–3, 6–7
Sales Data	4–3, 6–5
Sales Report	4–3
Sales Statistics	4–3
Selected Order Trans	6–8

4. Data Stores

Name	Figure
D1 Inventory	4–3, 6–5, 6–6, 6–7
D2 Master File	4–3, 6–5, 6–7, 6–8
D3 Sales File	4–3, 6–5, 6–7
D4 Customer File	4–3, 6–5, 6–7

5. Processes

Number	Figure
1–6	4–3
1.1–1.5	6–5
2.1, 2.2	6–6
4.1–4.5	6–7
5.1–5.3	6–8

For each of the above items, HPC makes a separate entry using the data dictionary formats (see Figures 5-11 through 5-13). For quick reference these are kept on 3 × 5 index cards.

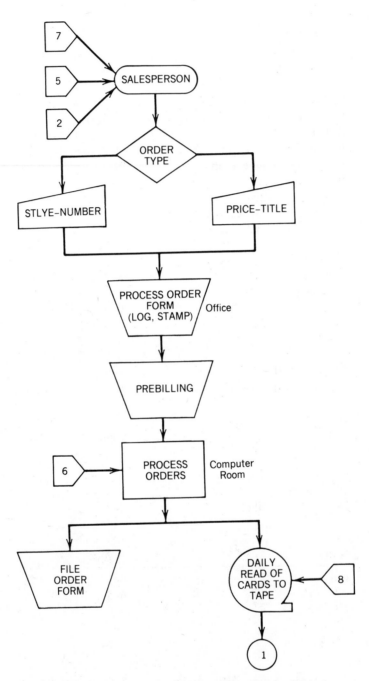

FIGURE 6-2   Flowchart of current system at Toy World.

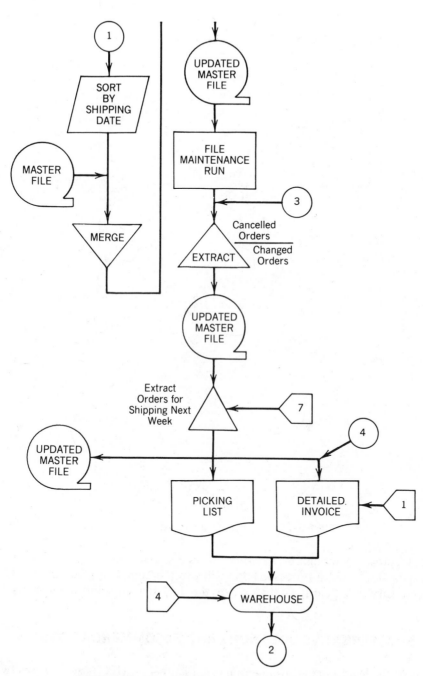

FIGURE 6-3 Flowchart of current system at Toy World.

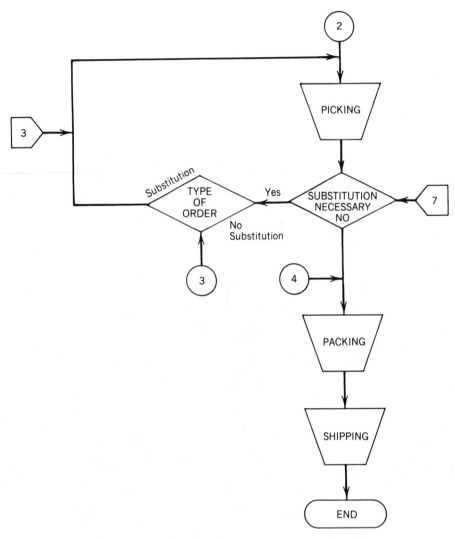

FIGURE 6-4 Flowchart of current system at Toy World.

Figures 6-9 through 6-11 represent three sample entries for a data flow, a data store, and a process. HPC will use the complete stack of the data dictionary index cards later to create the data elements and file formats.

6.5 ALTERNATIVE SOLUTIONS AND RECOMMENDATIONS

Using the level 1 data flow diagram (see Figure 4-3), HPC now prepares two solutions to the present problem. Both solutions are partially on-line and partially batch. In one option, process 1 is implemented on-line and all other

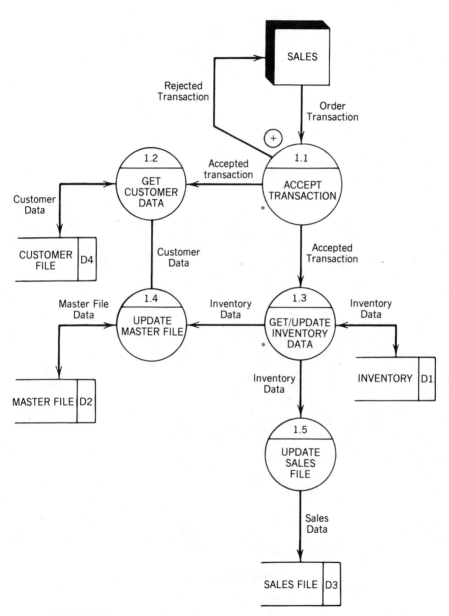

FIGURE 6-5 Level 2 data flow diagram for process 1 – PROCESS ORDER.

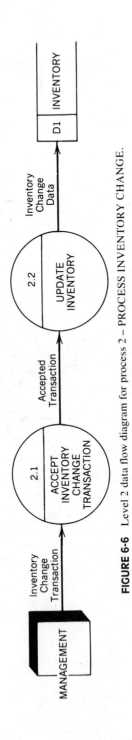

FIGURE 6-6 Level 2 data flow diagram for process 2 – PROCESS INVENTORY CHANGE.

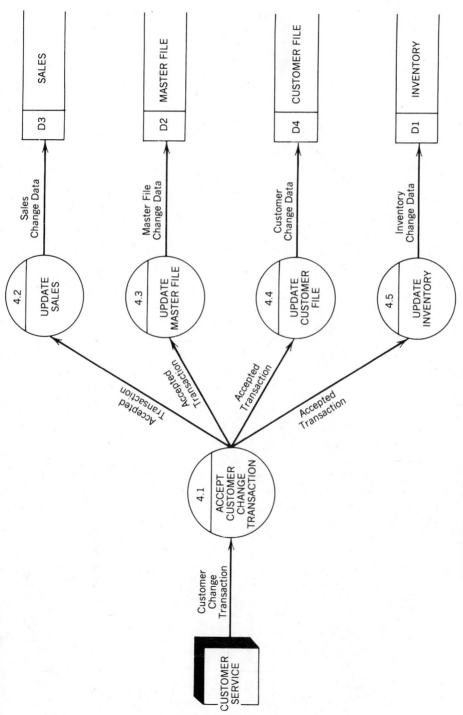

FIGURE 6-7 Level 2 data flow diagram for process 4 – PROCESS CUSTOMER CHANGE.

113

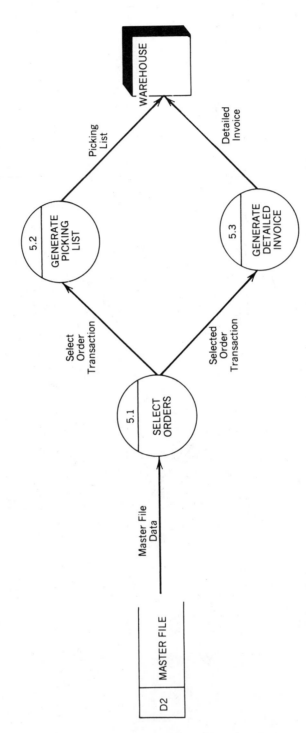

FIGURE 6-8 Level 2 data flow diagram for process 5 – PROCESS MASTER FILE.

Name: Sales Data

Aliases: None

Description: Data belonging to a sales receipt/order
             generated by process 1; e.g., invoice no.,
             item no., unit price, quantity sold, total
             price, date of sale.

Format: Not yet finalized

Location: Sales file (D3)

FIGURE 6-9   Data dictionary entry for data flow—Sales Data.

Name: Inventory

Aliases: D1

Description: Contains all data related to inventory;
             e.g., quantity on hand, item no., item
             description, unit price, reorder level,
             quantity on order.

FIGURE 6-10   Data dictionary entry for data store—Inventory.

Name: Accept Transaction

Number: 1.1

Description of Process Logic: Get and edit transaction.
   If transaction is good, then get customer data and
   get/update inventory data
   else reject transaction.

Input: Order Transaction received from Sales

Output: Accepted Transaction sent to processes 1.2 and
   1.3

FIGURE 6-11   Data dictionary entry for process 1.1—Accept transaction.

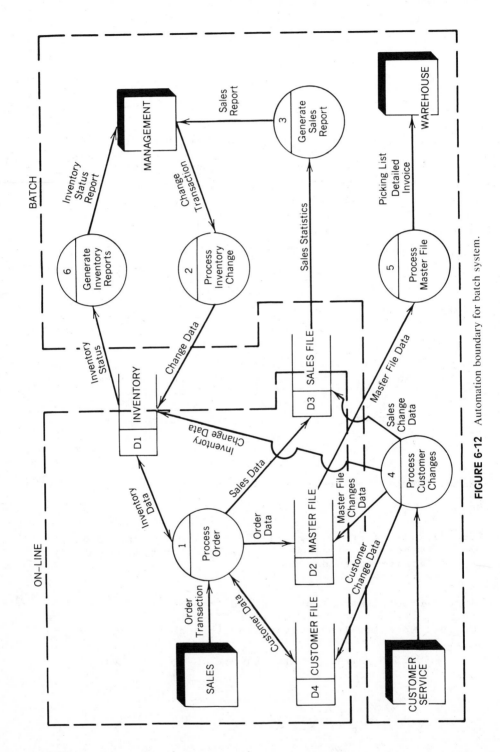

FIGURE 6-12 Automation boundary for batch system.

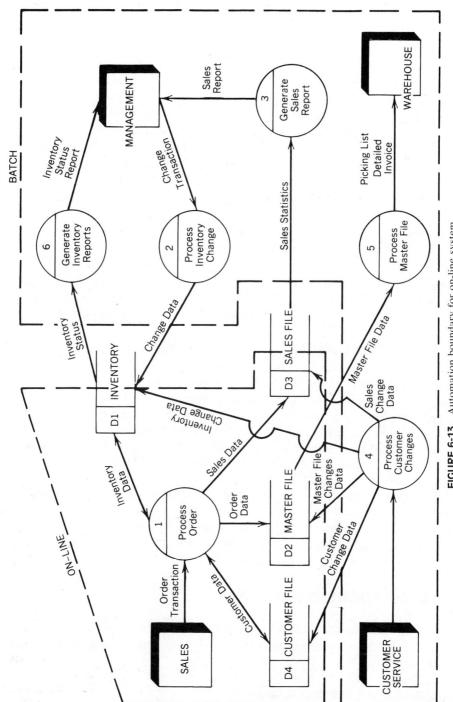

FIGURE 6-13 Automation boundary for on-line system.

	Alternative System (On-line)	Proposed System (Batch)
Economic feasibility	Ability to regain lost sales; i.e., 250,000 lost revenue, which is 2% of gross sales (14 million)	Ability to regain lost sales; i.e., 250,000 lost revenue, which is 2% of gross sales (14 million)
	Anticipate an additional 5% plus gain in sales due to better inventory control	Anticipate an additional 3% plus gain in sales due to better inventory control
	Estimated payback period 1½ years	Estimated payback period 1 year
Technical	System can be implemented using current technology	System can be implemented using current technology
Operational feasibility	Management indicates an in-house system would be operationally acceptable	Management indicates an in-house system would be operationally acceptable

FIGURE 6-14   Summary findings of feasibility issues.

processes are run batch. HPC calls this solution a batch system. In the other option, which HPC calls an on-line system, processes 1 and 4 are done on-line while the other four processes (i.e., 2, 3, 5, 6) are run as batch jobs. The time sensitivity used here pertains to the immediate processing of a sales transaction or both a sales transaction and a customer change request. Because of the size of the operation of Toy World, HPC recommends the batch version.

Figures 6-12 and 6-13 represent the two automation boundaries imposed on the level 1 data flow diagram in order to derive the two alternative solutions. Finally, Figure 6-14 gives the summary findings of HPC for the three feasibility issues related to both solutions. The payback periods are further substantiated via a cost/benefit analysis discussed in Section 6.6.

6.6   COST/BENEFIT ANALYSIS

HPC has prepared the following breakdown of time and cost for the analysis and design phases of the project. The total developmental cost then consists of two parts:

a. Analysis and design costs
b. Hardware and software costs

Item a is detailed as follows:

Step	Elapsed Time	Cost
Problem definition and feasibility study	Completed	$ 4,750
Analysis	3 weeks	5,500
System design	3 weeks	5,500
Detailed design	6 weeks	12,500
Implementation	6 weeks	13,750
Total	18 weeks	$42,000

Then the total developmental cost is given for two alternative solutions, the proposed system being the batch version:

	Proposed System (Batch)	Alternative System (On-line)
Analysis and Design	$ 42,000	$ 42,000
Hardware and Software	340,000	420,000
Total	$382,000	$462,000

The operating costs on an annual basis are as follows:

	Proposed System (Batch)	Alternative System (On-line)
Hardware and Software Maintenance	$48,000	$48,000
Utilities	3,500	6,000
Supplies	3,200	4,800
Total	$54,700	$58,800

As outlined in Figure 6-14, the total annual benefits are estimated as follows:

	Proposed System (Batch)	Alternative System (On-line)
Cash flow	$250,000	$250,000
Inventory Control	63,000	100,000
Staff reduction	64,000	124,000
Total	$377,000	$474,000

Assuming that Toy World accepts the recommendation of going with a batch system, the initial investment in the form of developmental costs is $382,000. This cost is compared against the annual net benefits, which equal the annual total benefits less annual operating costs, amounting to $322,300. Using the method of Section 5.9, case 2 and assuming an interest rate of 14 percent, the payback period of the batch system is 1.4 years, as shown below:

Year	Net Benefits	Interest	Present Value	Cumulative Present Value
1	$322,300	1.14	$282,719	$282,719
2	322,300	1.30	247,923	530,642
3	322,300	1.48	217,770	748,412

This is well within the financial constraints imposed by Toy World on the system project.

6.7 RECAPITULATION OF PROBLEM DEFINITION FOR FINANCIAL REPORTING SYSTEM

As described in Section 4.12, the financial reporting system (FRS) consists of four subsystems:

a. Data Entry/Edit/Display
b. Data Validation
c. Database Administration
d. Annual Report Generation

The proposed system will be predominantly menu driven and on-line. Subsystem (d), however, will not be a part of the system menu, since it will be run only once a year. Refer to Section 4.10 for more detailed descriptions of the subsystems.

6.8 DEFICIENCIES OF THE CURRENT REPORTING SYSTEM

HPC interviewed the director and the system manager of Massachusetts Educational Foundation in order to gather more information on the current system. Since the system manager is very particular about documenting the system, adequate materials such as system manuals, source code listings, system flowcharts, copies of data input sheets, and reports are available. HPC analyzed these documents very carefully, prepared a detailed system flowchart, and listed the following deficiencies and bottlenecks in the system.

6.8.1 Total Batch Operation

The system is run totally batch. As a result, there is no way to speed up those processes that are highly time sensitive. No capability exists to ask ad hoc questions and generate quick ad hoc reports. Often the director of the foundation is unable to answer questions from the press in a timely manner. If the answer is not available from the existing reports, the director has to wait until the necessary information is extracted from the system by the data processing staff.

6.8.2 Keypunch of Input Data

Data from the forms are transferred onto the input data sheets by the data analysts. These data sheets are then collected in batches and once a week are sent to a keypunch operation in Paramus, New Jersey, to be keypunched into cards. Cost is the only reason for hiring the out-of-state contractor to keypunch the data, since the transportation of the cards takes a long time and causes a severe bottleneck in the system. An average time between sending the input data sheets and receiving the keypunched card is 10 days.

6.8.3 Difficulty in Data Editing

Once data are entered into the database, it is very difficult to change them. This is due primarily to the batch nature of the system, although the design of the forms database is also partially responsible. In order to change an existing data value, the user must request a data change form and fill it out. These forms are collected and once a week are sent to the keypunch center in Paramus along with all other input data sheets. After the keypunched cards are back, the data edit program is run to change the data. As with data entry, data edit also has an average turnaround time of 10 days.

6.8.4 Faulty Database Design

For some obscure reason, the forms database was designed using the entity-relationship data model. Since there was no commercially available DBMS that could implement an entity-relationship model, a relational DBMS was used. To accommodate this situation, the data in each form were scattered over multiple entity relations and relationship relations. Consequently, in order to retrieve the complete data on a given form, it is now necessary to access three to five different relations. This makes the reconstruction of a form very difficult.

6.8.5 Data Changes Kept Manually

Due to legal requirements, complete records must be kept of all changes made to any data value in any form. The auditors examine these records on a random basis during the auditing. The current forms database does not

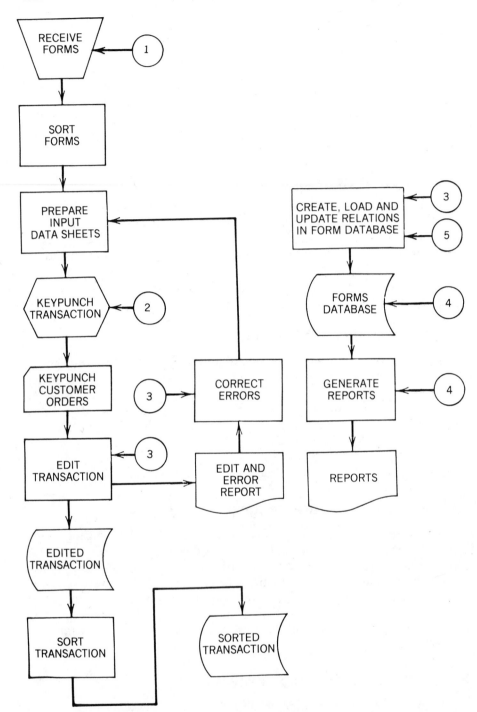

FIGURE 6-15 System flowchart for financial reporting system.

have any capability to record such change transactions in an automated form. Accordingly, all these records are kept manually as copies of data change sheets, letters from the grantees requesting the data changes, and hardcopies of the contents of the affected relations in the database. They are too bulky to store and it is too time consuming to retrieve them.

Figure 6-15 shows the system flowchart of the current system. The five deficiencies listed above are identified in the diagram using circled numbers 1 through 5.

6.9 EXPLOSIONS OF LEVEL 1 DATA FLOW DIAGRAM

In Section 4.13 the proposed logical system was described as consisting of five processes:

1. Entry/edit/display data
2. Update form data store
3. Validate entered and edited form
4. Query and Maintain form database
5. Generate annual report.

Figure 4-5 shows the level 1 data flow diagram for the system.

HPC now explodes each process into lower levels in order to analyze further the operation of the system. As samples of the exploded diagrams, HPC includes Figures 6-16 and 6-17 representing a three-level explosion of process 4, i.e., query and maintain form database.

6.10 SAMPLE DATA DICTIONARY ENTRIES

HPC has prepared an extensive list of all data flows, data stores, and level 3 processes for the financial reporting system. To avoid repetitions, three sample data dictionary entries, one each of a data flow, a data store, and a level 3 process, are given in Figures 6-18 through 6-20.

6.11 RECOMMENDED SOLUTION

The customer, Massachusetts Educational Foundation, already has an operational financial reporting system. Since it is completely in batch mode, a variety of system deficiencies and bottlenecks currently exist, as discussed in Section 6.8. Accordingly, the customer already decided on exactly what was wanted before hiring HPC. Therefore, during the initial meetings and again at the time of the more in-depth interviews during the system analysis

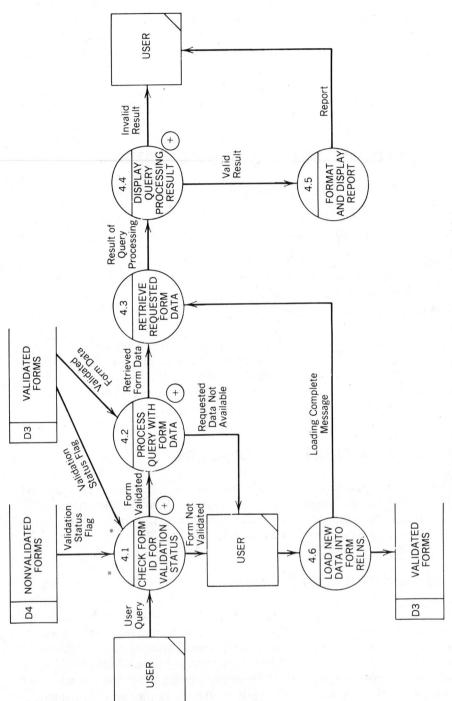

FIGURE 6-16 Data flow diagram for process 4 at level 2.

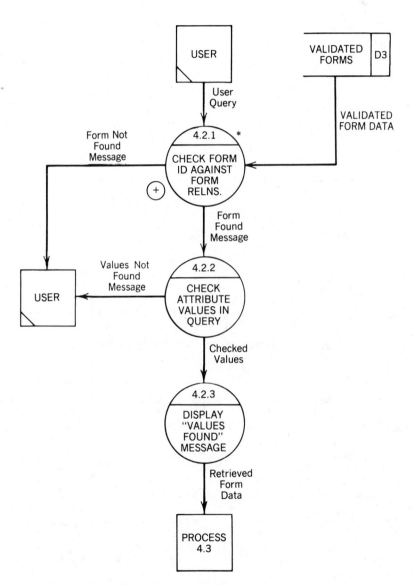

FIGURE 6-17   Data flow diagram for process 4.2 at level 3.

Name: Validated Form Data

Aliases: Valid Form, Validated Form

Description: Individual data elements belonging to any
             reporting form that has passed *all*
             validation checks in subsystem 3. Tentative
             data elements are form ID, grantee ID,
             valid status flag, and all relevant
             financial data.

Format: To be decided

Location: Validated Forms (D3)

FIGURE 6-18   Data dictionary entry for data flow—Validated Form Data.

Name: Validated Forms

Aliases: D3

Description: Contains records on forms validated via
             subsystem 3, Each record has data elements
             on form ID, grantee ID, valid status flag,
             and other pertinent financial data.

FIGURE 6-19   Data dictionary entry for data store—Validated Forms.

Name: Check Form ID against Form Relations

Number: 4.2.1

Description of Process Logic: Read form ID and grantee
    ID from User Query. Search data store, Validated
    Forms, for the records matching the form ID and the
    grantee ID. If a match is found, send ``found''
    message to process 4.2.2
    else display ``not found'' message to user.

Input: User Query received from User Validated Form Data
       received from data store, Validated Forms (D3)

Output: Form Found Message sent to process 4.2.2; Form
        Not Found Message displayed to the User

FIGURE 6-20   Data dictionary entry for process 4.2.1—Check Form ID against Form Relations.

phase, the director and the system manager of the foundation told HPC that they did not want a set of two or more alternative solutions to their problem. Instead, they wanted only one recommended system, partially on-line and partially batch, that would meet the following objectives:

a. Data entry/edit/display will be completely on-line.

b. Data storage in and data retrieval from the forms database will be both batch and on-line.

c. Query report generation will be totally on-line.

d. Data validation will be both batch and on-line.

e. Generation of prescheduled periodic reports will be completely batch.

Because the customer has already grouped the various processes into separate categories based on their sensitivity to response times, HPC does not think it necessary to impose alternative automation boundaries on the level 1 data flow diagram. Thus, HPC provides the following detailed description of their recommended solution.

The on-line part of the system will be menu-driven with two categories of users. One category will be called USER and will have access to the following subsystems (see Sections 4.11 and 4.12):

a. Data entry/edit/display

b. Data validation

The other category will be called ADMIN and will have access to the following subsystems (see Sections 4.11 and 4.12):

a. Data entry/edit/display

b. Data validation

c. Database administration

In general, on a regular basis (say, daily during peak season and weekly during slack season) the forms entered during the period will be first validated and then stored in the forms database. This operation will occur in batch mode. However, if a user wants to make small edits or validations to a few forms, he or she will be able to do so via the menu options. This will provide the on-line version of data validation and data storage/retrieval from the forms database.

An authorized user, i.e., one classified as ADMIN, will be able to query and maintain the forms database on-line via the menu option of database administration. However, the regular updating of the database will be done in batch mode. Finally, all prescheduled periodic reports and the annual report will be generated in batch mode alone.

6.12 COST ESTIMATES

The customer realizes that the proposed system will not be cost justifiable in terms of hard dollars. In other words, there will not be enough tangible benefits to offset the developmental cost and yield a finite payback period. This is due to the fact that the customer currently has an operational batch system that meets the same objectives albeit in a slow and error-prone way. Accordingly, the benefits accruing from the new system will be predominantly intangible, such as greater satisfaction and improved morale of the employees, better record tracking capabilities, and improved public relations. So the customer already told HPC not to attempt a formal cost/benefit analysis but only to provide a detailed cost estimate of the project.

HPC now examines the two types of cost: developmental and maintenance. Since the new system will replace the current batch system, there will be no appreciable change in the maintenance cost. Since the customer's annual budget already covers the maintenance cost of the batch system, HPC decides to estimate only the developmental cost of the new system.

The customer already owns the hardware, which is a PDP-11/70 with 2.0 MB of memory and a UNIX/5 operating system. Also, since the entire operation will be performed in a single location, no data communication costs are involved. HPC itemizes the developmental costs as follows:

Additional Hardware (e.g., CRT terminals, disk storage, printers)	$ 20,000
Personnel (e.g., HPC team, customer's employees assigned to the project)	270,000
Total	$290,000

Of the total personnel cost of $270,000, it is estimated that $80,000 will be allocated for the customer's employees assigned to the project and the remaining $190,000 will be expended by the HPC team. The personnel cost for HPC team members is further detailed as follows:

Phase	Cost
Problem Definition and Feasibility	$ 20,000
System Analysis	20,000
Preliminary System Design	30,000
Detailed System Design	60,000
System Implementation and Evaluation	60,000
Total	$190,000

Thus, the total developmental cost is estimated to be $290,000.

6.13 SUMMARY

This chapter continued the two case studies, introduced in Chapter 4, through the detailed system analysis phase.

The problem with Toy World, Inc., lies in the order processing, inventory control, and sales management areas. The system consultant HPC looked into the existing system and identified eight deficiencies and bottlenecks that plagued the system. Part of the blame was the manual nature of the current operations and part was the managerial decision-making process, which was not always supported by necessary data. HPC next exploded the level 1 data flow diagram of Chapter 4 (see Figure 4-3) into a set of more detailed level 2 diagrams. From these a complete set of data flow, data store, and process entries for the data dictionary were made. One sample of each entry was provided. Afterward, HPC imposed two alternative automation boundaries on the level 1 data flow diagram and came up with two solutions, one being predominantly batch and the other primarily on-line. Since the operations of Toy World are not critically time sensitive, HPC recommended the predominantly batch system.

The final task item of HPC was to prepare a detailed cost/benefit analysis. It was estimated that the batch system would incur a developmental cost of $382,000 and an annual benefit of $322,300. This yielded a payback period of 1.4 years. These findings met the approval of Toy World management, and HPC proceeded to the next phase of preliminary system design.

The system for Massachusetts Educational Foundation deals with financial reporting. Their current operation is completely in batch mode. As with the study for Toy World, HPC first identified five deficiencies of the current system. The main problem was the batch mode of operation, which slows things down and makes data editing difficult. HPC next prepared exploded data flow diagrams up to level 3. Samples of these diagrams were shown in the chapter. Then HPC prepared a detailed data dictionary for the system, including a separate entry for each data flow, data store, and process.

Since the foundation told HPC about the exact requirements for the new system, HPC did not provide alternative solutions. Instead, it gave a detailed description of a proposed system that would meet the customer's needs. Knowing that most of the benefits from the new system would be of an intangible nature, HPC did not compute any payback period but merely estimated the developmental cost as $290,000.

6.14 KEY WORDS

The following key words are used in this chapter:

alternative solutions   exploded data flow diagram
automation boundary   inventory control

cost/benefit analysis

data dictionary

data flow

data store

data validation

database

destination

developmental cost

operating cost

order processing

payback period

process

sales management

source

system deficiencies

system flowchart

REVIEW QUESTIONS

1. Describe the functional capabilities of the system proposed for Toy World.
2. Explain why the eight items listed as system deficiencies for Toy World cause some of its current problems.
3. Select two data flows from Section 6.4 and prepare data dictionary entries for them.
4. Why does HPC recommend a primarily batch system for Toy World?
5. Compute the payback period for the on-line option of the system proposed for Toy World (see Section 6.6).
6. The batch mode of operation is at the source of all problems with the financial reporting system—justify or disprove (see Section 6.8).
7. Why does HPC not provide multiple alternative solutions for the financial reporting system?
8. Assuming an annual net benefit of $45,000 for the financial reporting system, compute the payback period.

STRUCTURED DESIGN

Part III consists of Chapters 7 through 10. Chapter 7 discusses the theory of preliminary system design, which is the third phase of the system life cycle. It describes the transition from the analysis to the design phase so that the emphasis now is on how the capabilities of the proposed system are going to be implemented. In this phase, the input and the output are described in detail, while an overview is provided of the processing part of the proposed system. Chapter 8 then continues the two case studies through the preliminary system design phase. Chapter 9 describes the detailed system design, the fourth phase of the system life cycle. It addresses the topics of file access, file structuring, estimate of auxiliary storage, overview of data communication network, equipment specification, and personnel selection for system implementation and maintenance. A brief mention is made of the schema structuring of a database. Chapter 10 carries out the detailed design for the two case studies.

Preliminary
System Design:
Theory

7.1 BASICS OF SYSTEM DESIGN

The first two phases of system development address the *what* aspect of the proposed system. As noted in Section 5.10, the end product of these two phases is a complete specification of the logical system consisting of the data-flow diagrams, data dictionary entries, and a description of the subsystems and their interfaces. The third and the fourth phases consist of the conversion of the logical system into a physical system. They address the *how* aspect of the new system, i.e., *how* will the component subsystems be implemented. The transition from *what* to *how* marks the beginning of the system design phases.

The physical system is implementation bound. Consequently, the system flowchart is used as the graphic tool to describe the system in the design phases. The data flow diagrams prepared during the analysis phases are now converted into corresponding system flowcharts. The logical processes now become would-be programs that can implement the functions of the processes. Similar conversions are done to the data flows and data stores (see Section 7.9).

The system design process can be regarded as a black box approach to the new physical system. As illustrated in Figure 7-1, the new system appears as a black box that accepts the input data and produces the output information

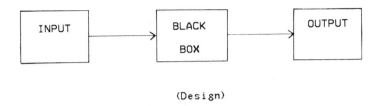

(Design)

FIGURE 7-1 Black box approach to system design.

as reports and/or screen displays. The user specifies the output needed from the system. The system team matches the output with appropriate input, and the black box provides the necessary link between the input and the output.

7.2 TWO PHASES OF SYSTEM DESIGN

The system design process is split into two phases:

a. Preliminary design
b. Detailed design

The distinction between these two phases is one of degree only, not of kind. In this respect, the situation is similar to the first two phases of system development; i.e., problem definition and feasibility, and system analysis. Although there is general agreement regarding the issues addressed in the preliminary and detailed design phases, the grouping of these issues into two separate phases is somewhat arbitrary. Nevertheless, in this book we shall include the following topics in preliminary design:

☐ Detailed layout of the output requirements
☐ Detailed layout of the input documents
☐ An overview of the processing component of the system

The detailed design will expand on the last item by addressing the system level design and related issues.

User involvement in all phases of system development is the basic premise of structured techniques. As a result, user input plays a major role in the analysis phases. The same principle applies to the design phases. Since the user is primarily interested in the output of the system, the preliminary system design starts with a detailed description of all the output that the system produces for the user.

7.3 TWO TYPES OF OUTPUT

In general, there are two types of output:

a. Printed report
b. Screen display

For a purely batch system, the output consists entirely of printed reports. For on-line systems, the screen displays play a significant role. The output from the system is the primary contact between the system and its users. The quality of the output and its usefulness are two major factors in determining whether the system will succeed. Consequently, it is essential to have the best possible form of output.

If the output is a printed report, proper care should be taken for its format, such as heading, spacing, subtotals, and grand totals. If it is a multipage report, the correct page number must appear on each page, and all "constant" information such as the title of the report, column names, and the date of the report must be repeated on each page. If any codes are used in the main body of the report, a separate explanatory paragraph (e.g., legend) must be printed at the bottom of each page of the report. Wetherbe ([6], pp. 120–127) has provided a detailed discussion of output definitions for printed reports.

The format and appearance of screen displays are more complex to handle. The primary reason for the complexity is the level of human interaction involved in screen formatting. Since on-line systems are now becoming increasingly popular, screen displays form a very important part of the system output. We shall discuss these issues in more details in Sections 7.5 and 7.6.

7.4 THREE TYPES OF REPORT

There are three types of report that an organization normally uses in its daily operation:

a. Detailed
b. Summary
c. Exception

These three types are geared toward different types of users and the levels of their involvement.

A *detailed report* is a complete tabulation of all the items and/or values pertaining to a specific area. For example, listings of all the items in the inventory, or all employees in a company, or all customers of a firm, and so on, are detailed reports. They are used primarily by the operational level employees and first-line supervisors in order to perform the daily business chores. If a detailed report includes multiple areas, then a control break appears at the end of each area. Thus, a master list of all employees in a company with multiple departments such as accounting, production, and information system should contain separate breakdowns by department with appropriate subtotals such as the total number of employees in a department and total salary paid to all the employees in a department.

A *summary report* is a subtotal of detail by categories in a detailed report. Thus, a row in a summary report may represent a category subtotal from a detailed report. For instance, an individual row in a sales summary report for a multistore company may consist of the location, the manager's name, and the total sales amount for a store. The main users of summary reports are middle level managers who want to supervise the operations of several departments, stores, areas, and so on, or want to compare the performance of one department with that of another.

An *exception report* isolates detailed or summary information from a detailed report or a summary report, respectively, in order to emphasize certain information that significantly deviates from the normal. For example, an exception report generated by an inventory control system may contain a listing of all inventory items for which purchase requisitions should be made since their present inventory levels have fallen below their reorder levels. Similarly, an exception report from a sales processing system may contain a listing of the names, store locations, and total bonus amounts for all those salespeople whose bonuses during the latest reporting period have fallen below the minimum limit established by the company. Thus, exception reports are used to set up some kind of flags or warning signals related to the daily operation of the organization. Main users of these reports are middle and upper levels of managers.

Both summary and exception reports are powerful tools of an information system. They highlight and isolate important detailed information. This saves a considerable amount of time for the decision-making management. The system team must determine what kinds of reports are required by which users to formulate the output requirements for the preliminary system design phase.

The design team should stress clarity in format and headings for output; one of the most frequent complaints is that users cannot understand reports. It is essential to use clear, descriptive titles for different fields on the report and avoid the use of obscure or little-known codes. If the data on a report are not obvious, use a footnote explaining how the numbers were derived, or refer users to the appropriate user documentation for an explanation. Consider including a short description of how the computations were performed on the first page of the report.

When it is necessary to print exception or error messages, the system should provide as polite a response as possible. Here again, it is not desirable to use coded numbers such as error 13B. Instead, explain what is wrong and

Symbol Code	Symbol Description
A	Alphabetic
X	Alphanumeric; i.e., digits or characters
Z	Zero-suppressed numeric field
9	Numeric field
B	Embedded blank in an alphanumeric field
.	Decimal point
$	Floating dollar sign

FIGURE 7-2 Standard symbols used for report formatting.

try to have the error message provide feedback so the user will learn from such mistakes.

The printer spacing chart is the standard tool used for designing a report. A report consists of two main parts: headers and information. Headers are the report titles, column headings, page numbers, and dates associated with a report or display. Information is the actual content of the output.

Headers may be printed by the computer or preprinted on custom-made forms. In either case, the headers are formatted on the layout form by positioning the letters of the headers using column and line numbers.

After the headers are formatted, the format of the information to be printed under the headers can be laid out. Figure 7-2 shows the symbols used to convey, in a standard fashion, the forms in which information fields are to appear on the report. Information fields that are printed on successive lines need only have print symbols shown for the first line. A wavy vertical line drawn under a field is sufficient to convey that the field is to be repeated on successive lines (see Figure 7-3).

The conditions causing each line of a report or a display to be printed should also be conveyed on the report layout. The condition causing a line to be printed is generally a change in the values of certain information fields or the start of a new page.

Figures 7-3 and 7-4 show examples of sample report formats using the printer spacing chart. Figure 7-3 is a detailed report, and Figure 7-4 is a summary report. Looking at Figure 7-3, we find that it has four separate header lines: The first header line gives the date, the name of the report, and the page number; the second header line gives the report period; the third header contains the name of the organization; the fourth header line gives the column titles of the seven columns. All these header lines are repeated on each page of the report. The wavy vertical lines indicate repetitions of detailed rows of data under each column title. When a department changes, two subtotals are printed with titles:

```
ACCOUNT TOTALS
ACCOUNT GROUP TOTALS.
```

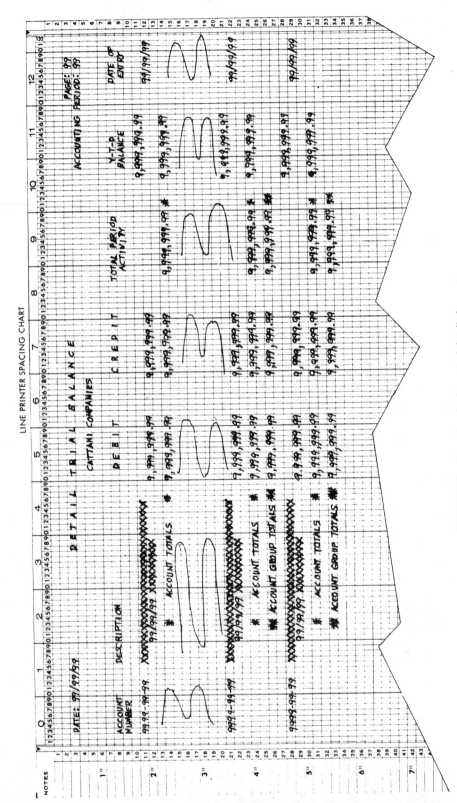

FIGURE 7-3 Example of a formatted detailed report.

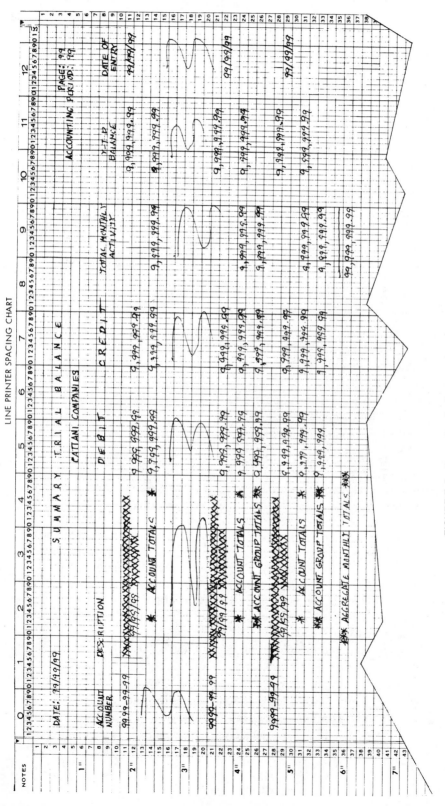

FIGURE 7-4 Example of a formatted summary report.

The standard symbols given in Figure 7-2 have been used to format the report in Figure 7-3. Similar comments apply to Figure 7-4.

7.5 SCREEN DISPLAYS

In Section 7.3 we mentioned the screen displays as a second type of output. They play a significant role in on-line systems. The user accesses the system, enters and edits data, updates the files, queries the database, and generates reports in an interactive mode. As a result, an extra dimension must be added to screen formatting to ensure that system prompts and responses are user-friendly. Clearly, what is user-friendly for an experienced system user is normally *not* user-friendly for a novice. Hence, there should exist multiple options and HELP information at various levels to address the needs of different groups of users.

Screens are mostly designed in a menu-driven form, which works as follows:

- **a.** The user logs in and is greeted by the system with a welcome message.
- **b.** The system displays the main menu, which shows the list of options available.
- **c.** The user selects an option and types in its code, which may be a number or a mnemonic.
- **d.** The system responds by performing the requested function.
- **e.** The user exits from the system by logging out.

Steps (c) and (d) are essentially iterative in nature since the user may have multiple requests or the system may go into multiple-level detailed menus from the main menu to respond to the user's questions or commands.

Figures 7-5 and 7-6 show examples of two menus from an on-line system called Investment Decision Support Systems, or IDSS. The user can go from the main menu in Figure 7-5 to the detailed menu in Figure 7-6 by selecting option 1 in Figure 7-5. Figure 7-6 prompts the user for necessary information. The human interaction with the system is the prime concern in screen designs. System efficiency may even be compromised to make the screen design more responsive to user needs.

On-line transactions using screens are much more efficient than the older batch operations. The advantages are as follows:

- **a.** The data entry process is faster since the user can correct errors by means of the dialogue with the computer.
- **b.** The data edit process is accomplished quickly by allowing the user to reenter data in order to rectify errors.

IDSS

MAIN MENU

IDSS can be used by management consultants and portfolio managers to maximize efficiency in investment decision making. It is suitable for both short- and long-term decision making. Available options are;

1.   CLIENT PORTFOLIO INFORMATION
2.   SECURITY SELECTION ANALYSIS
3.   INVESTMENT TIMING ANALYSIS
4.   PORTFOLIO ANALYSIS TECHNIQUES
E   EXIT FROM SYSTEM
H   HELP INFORMATION

SELECT OPTION AND PRESS RETURN KEY.

FIGURE 7-5 Sample screen for main menu. (Reproduced by permission from S.S. Mittra, *Decision Support Systems*, Wiley, New York, 1986.)

IDSS

CLIENT PORTFOLIO INFORMATION

Investment Type
Codes are

CLIENT NAME ?   1 - Common Stocks
INVESTMENT TYPE CODE ?   2 - Preferred Stocks
EARNINGS PER SHARE REPORT (Y/N) ?   3 - Bonds
DETAILED INVESTMENT REPORT (Y/N) ?   4 - Money Market
NEWS REPORT (Y/N) ?   Mutual Funds
PRESS RETURN KEY .   5 - Other Mutual Funds
                                            6 - Commercial Paper
                                            7 - Others
                                            8 - All Investments

FIGURE 7-6 Sample screen for a detailed menu. Reproduced with permission from S.S. Mittra, *Decision Support Systems*, Wiley, New York, 1986.

   c. Simple ad hoc reports are generated easily without using a full-fledged application program.

   d. The debugging phase is speeded up for programmers.

7.6 HUMAN FACTORS IN SCREEN DESIGN

In 1960 Calvin Mooers proposed the following law, known as *Mooers' Law*, to emphasize the necessity for addressing human factors in general on-line systems ([2], Chapter 17):

> *Mooers' Law*—An information retrieval system will tend *not* to be used whenever it is more painful and troublesome for a customer to have information than for him not to have it.

Later in 1971 W. J. Hansen enunciated the three principles known as *Hansen's user engineering principles* ([2], Chapter 17):

☐ Minimize memorization
☐ Optimize operations
☐ Engineer for errors

James Martin ([4], Chapter 17) has commented extensively on the *three levels* at which attention should be paid to user psychology:

   1. Functional Level. *What* functions should be performed by people and *what* by machine? This question relates to the *analysis phase*. People should be able to

     ☐ Act flexibly and intelligently when exceptional situation arises.

     ☐ Deal with ambiguities in language or data.

     ☐ Deal with errors usually caused by data entry. Error messages should be brief and clear. Corrections should be least disruptive to the data entry process. The user normally faces two types of errors during any interactive session: typing errors and comprehension errors (relating to system procedures and protocols).

   2. Procedural Level. *How* to implement the capabilities derived at functional level. This level relates to the *design phase*. Factors to be considered are

     ☐ Hardware: Type and location of terminals, keyboard with keys dedicated to major system commands, visual quality of screen, size of characters displayed (e.g., Dec Writer III allows the user to select 10 to 16 characters to an inch).

☐ Language of communication: Not too many messages displayed at the same time, dialogue be clear and simple, preferably menu driven.

☐ Conflicting user needs: Cater to the needs of beginning and advanced users; e.g., provide abbreviations of commands for advanced users.

☐ User acceptance: Beginners are often intimidated by immediacy of feedback, lack of privacy in a terminal cluster area, phobia relating to the fragility of the system. Use training to dispell such fears.

☐ Provide an EXIT at each frame displayed.

3. Syntactical Level. *How* the design can be best implemented. This level relates to the *implementation phase*. Issues to be considered are

a. *System Sensitivity*

☐ Display a small amount of information at a time.

☐ Have one idea per display.

☐ Always let the computer respond to the user (either as desired info or as signal to wait).

b. *Display Characteristics*

☐ Make instructions to the operator stand out.

☐ Clear up the screen when possible.

☐ Make HELP easily available.

c. *Availability Factors*

☐ Make system easily available to a large number of users.

☐ Balance trade-off; i.e., neither a too overloaded system all the time so that user cannot login nor a too restrictive user group so that system is not popular.

7.7 ISSUES INVOLVING HUMAN INTERACTION IN SCREEN DESIGN

We shall address the following issues in designing screens:

a. Design consistency

b. Dialogue design

c. Performance criteria

d. Bullet proofing

e. Screen formatting software

7.7.1 Design Consistency

Design consistency ensures that the system provides prompts and messages in a uniform manner to help the user. Thereby it reduces the learning time for the user. Consistency must be achieved in keying procedures, screen format, display area of prompts and messages, and the language used in prompts, messages, and "help" information. Thus, each time the user hits a key, the same predictable action should take place. A designated area on the screen, say the top two lines or the bottom two lines, should be reserved for displaying system messages such as error messages or prompts. The language used should preferably be in the active voice and second person. For example, a message should appear as

```
PRESS RETURN AFTER YOU ARE DONE
```

instead of

```
RETURN KEY SHOULD BE PRESSED AT END
```

If the system sounds "beep" at one error, then the same beep for the same length of time must be sounded at every comparable error. The whole idea behind design consistency is to let the user build up some expectations and then satisfy that expectation consistently throughout any session.

7.7.2 Dialogue Design

The user interacts with the system in a two-way dialogue via the terminal. A prompt or a message is a dialogue between the user and the system. Help information can be regarded as a modified dialogue in the sense that it consists of some technical information followed by a direction for the user either to ask for more help information or to quit the help session. James Martin ([4], p. 39) summarized the main issues in dialogue design as follows:

> We need to design the dialogue structure so that it is as easy as possible for humans to use. The degree to which we achieve this will determine the acceptability of the systems we create. However, the more freedom we give the operator in his phrase structuring the more complexity will be necessary in programming the computer to interpret what he says.

Figure 7-7 gives a list of steps to be used for dialogue design. Martin ([4], Chapter 2) has given a more detailed list of such steps. Lucas ([3], pp. 274–278) has used these steps for dialogue design.

We now describe a variety of dialogues that can take place between a user and the system. The discussion is adapted from Martin's treatment ([4], Chapter 7):

1. **Simple query.** The user enters a sequence of digits or characters (e.g., a part number or an employee last name), and the screen displays the associated data (e.g., part description or employee full name).

1. Simple query
2. Mnemonics
3. English language query
4. Function keys
5. Building a record on the screen
6. Operator instruction
7. Menu selection
8. Use of displayed format
9. Text editing
10. Hybrid techniques

FIGURE 7-7 Types of dialogues between computer and user.

2. **Mnemonics.** We can construct abbreviations that are meaningful to the user but require less keying than full text.

3. **English language query.** The user enters in plain English a full descriptive clause or sentence indicating his or her request.

4. **Function keys.** One or more function keys can be used to implement a set of user requirements such as printing a list or displaying a record on the screen.

5. **Building a record on the screen.** The user enters one or more commands or data elements, and the system builds a complete record such as employee information on the screen.

6. **Operator instruction.** The screen displays instructions such as "Enter Y or N" or "Press Return," and the user follows them to continue the dialogue.

7. **Menu selection.** The screen displays a variety of numbered options, say. The user enters a number to call the corresponding option. Alternatively, the user may move the cursor to the desired option and then press Return to call the option. The menu mode is very popular and often involves multiple levels whereby the user moves from one level to the next by responding to prompts.

8. **Use of displayed format.** Using displayed formats is similar to a fill-in-the-blank form, whereby the user writes over the blanks and erases the format during data entry. For example, a format MM/DD/YY may be displayed, and the user enters the values in the appropriate slots. This technique can be extended to a form filling format.

9. **Text editing.** The text editor allows the user to insert and delete characters, change character strings and numbers, move lines to other parts of the text, search for character strings in the text, and so on.

10. Hybrid techniques. Hybrid techniques involve interactive communication with the program. The conversation is initiated partly by the user and partly by the program.

7.7.3 Performance Criteria

Performance criteria establish some kind of quantitative measures to judge the quality of system responses to the user demands. The following are commonly used performance criteria:

   i. Response time
  ii. Form of system output
 iii. Unit cost as a performance measure

i. Response Time. Response time is the interval between the user's pressing the last key in the input operation and the terminal's displaying or typing the first character of response. An average response time of 10 seconds or less is regarded as acceptable. However, the response time for complex queries of a large database may be higher. In that case, the system should display messages indicating that it is still searching the database.

Several studies have been done on the mean and standard deviation of response times from different interactive computer systems. In general, a too high standard deviation on response time is undesirable. It causes inconsistency for user expectations of response times. Martin ([4], p. 323) has given the following rule of thumb:

The standard deviation σ of response time should not be more than half the mean μ of response time; that is

$$\sigma \le \frac{1}{2} \mu$$

Fast response time is

☐ Necessary for an acceptable and efficient human–machine interaction
☐ Conducive to economic savings
☐ Impressive as "public relations"

Too short a response time may be harassing for a slow-thinking operator.

In 1971, Higgins and Smith listed the following factors to achieve a fast response time ([2], Chapter 6):

☐ Speed of computer
☐ Core storage available

- ☐ Operating system
- ☐ Number of users simultaneously on-line
- ☐ Efficiency of the program

ii. Form of System Output. The system output provided as screen displays should

- ☐ Be clear and legible
- ☐ Have prompts and messages in reverse video or boldface
- ☐ Have a blinking cursor to attract the user's attention
- ☐ Have messages appearing at the same area of the screen, top, bottom, etc.

iii. Unit Cost as Performance Measure. A long response time keeps the user waiting, and an economic value may be assigned as the waiting cost, a term widely used in queueing theory. Users feel frustrated with systems having a high waiting cost.

The *efficiency ratio* is sometimes used as a quantitative measure to assess the system's performance:

$$\text{Efficiency ratio (ER)} = \text{mean number of lines entered per unit time (e.g.,} \\ \text{5 minutes, 10 minutes)}$$

ER may be misleading since it depends on the experience of the user.

7.7.4 Bullet Proofing

Bullet proofing protects a system from misuse by users and does not confuse users with imprecise or misguided dialogue. Steps to be taken for bullet proofing a dialogue are as follows ([4], Chapter 31):

i. Nonautomated checking—two people play the session, with one being user and the other the computer
ii. Live operation with selected operators
iii. Dialogue control group—specially needed for commercial applications, similar to quality assurance group for programming
iv. Training users

7.7.5 Screen Formatting Software

A variety of software products is available to design and implement screens. They can be used to generate forms interactively and then store them in files or databases, to set up multiple-level menus with associated HELP files, and for other similar purposes. Examples of screen formatting packages are

☐ THE MENU SYSTEM   Vendor: Soft Test Inc.
☐ CFI Operating Menu   Vendor: Consumers Financial Institute
☐ SCOPE   Vendor: Software House
☐ FMS Vendor: Digital Equipment Corporation
☐ SCP22   Available from Transportation Systems Center

In addition, many DBMS packages contain special application builders that can be used to prepare menus.

Figure 7-8 is an example of interactive menu produced by the CFI Operating Menu. Figures 7-9 and 7-10 show two data entry forms created by FMS and SCP22, respectively. Using the interface between FMS and DATATRIEVE, it is possible to access and manipulate the data entered via forms created with FMS. Similarly, using system 1022 (a DBMS available

```
        CFI Operating Menu

     SECTION 15 REPORTING SYSTEM

    =====> Forms Entry Program
           Echo Report Generation
           Exit




        Hit RETURN to advance the arrow, SPACE to select, ? for help

  Forms Entry Program ...
  THE DEFAULT NUMBER MAY BE ENTERED BY PRESSING THE RETURN KEY
  please enter the two digit fiscal year (default = 83):
  please enter ID (default = 3010):
  please enter the level which you are reporting (default = b):
  please enter a form number:

            Fig. 1  Parameter Collection without purchased of direct option.



  Forms Entry Program ...
  THE DEFAULT NUMBER MAY BE ENTERED BY PRESSING THE RETURN KEY
  please enter the two digit fiscal year (default = 83):
  please enter ID (default = 3010):
  please enter the level which you are reporting (default = b):
  please enter a form number: 407
  please enter p for purchased of d for direct:

            Fig. 2  Parameter collection with purchased of direct option.
```

FIGURE 7-8   Example of a CFI generated menu.

REVENUE SUMMARY SCHEDULE

| REVENUE OBJECT | TOTAL |
CLASS	REVENUE
PASSENGER FARES FOR TRANSIT SERVICE	$ +---------------
	+---------------
SPECIAL TRANSIT FARES	+---------------
	+---------------
SCHOOL BUS SERVICE REVENUES	+---------------
	+---------------
FREIGHT TARIFFS	+---------------
	+---------------
CHARTER SERVICE REVENUES	+---------------
	+---------------

FIGURE 7-9   Example of a data entry form created by FMS.

on DEC-10) and SCP22, you can generate reports with data entered by forms generated by SCP22.

W. O. Galitz has provided a 10-step checklist to design screens ([1], pp. 169):

a. Review screen design documentation and services.
b. Identify system input and output.
c. Identify unique user requirements.
d. Describe data elements.
e. Develop transactions.
f. Develop final paper screens.
g. Define computer screens.
h. Test screens.
i. Implement screens.
j. Evaluate screens.

```
            AIRCRAFT  ACCIDENT/INCIDENT  PRELIMINARY  NOTICE

FROM  _____      TO  _____

DATE  _____       TIME  ____

A 1      INFORMATION FROM  _____

B 1      REGISTRATION NO.  _____      B 2  MAKE/MODEL  _!_!_____
                                         (ENTER SECOND CHARACTER AS  /  OR  - )

B 3      OPERATOR OF AIRCRAFT  _____

B 4      TYPE  OF  ACTIVITY  _____

B 5      BRIEF DESCRIPTION  _____;
                            _____;
                            _____;
                            _____

B 6      WEATHER DATA  _____

B 7      AIRCRAFT DAMAGE :      A = DESTROYED
                                B = SUBSTANTIAL
                                C = MINOR
                                D = FIRE
                                E = NONE
                                   (ENTER A,B,C,D, OR E)

C 1      NAME/ADDRESS OF PILOT/INJURY  _____

C 2      NAMES OF CREW/INJURY  _____

C 3      NO. OF PASSENGERS/INJURIES  _____

D 1      LOCATION OF OCCURRENCE  _____
```

FIGURE 7-10   Example of a data entry form created by SCP22.

7.8 INPUT DOCUMENTS AND DATA ENTRY

Input documents provide data into the system. The data must be designed so as to match the output to be generated by the system. The input data come from a variety of source documents. The recent trend is to capture data as close to the point of generation as possible. This leads to less errors in the data entry process. Typical source documents are customer order forms, employee time cards, vendor invoices, purchase orders, and sales receipts.

Depending on the structure of the data files used in the system, two situations can arise:

a. The record format is very similar to the structure of one or more source documents.

b. The record format does not closely match any single source document.

In case (a), data may be entered directly from the source documents into the system. The record format should be designed so as to allow the data entry to proceed from left to right and then from top to bottom within a given source document. In case (b), a separate data entry sheet should be prepared to match an individual record format. Someone must manually complete a data entry sheet from the corresponding source documents before entering data into the system. In this situation, usually two types of people are involved—data analysts and data entry operators. The data analyst should prepare the data entry sheet; then the data entry operator should enter data from the completed sheet. The process of having two different people prepare and enter the data is usually less prone to error.

Input data may be entered in batch mode or on-line. Batch input suffers from its rigidity, since most batch input requires information in a strict format. Also, there is a lack of immediate feedback on errors. However, batch entry of data is less expensive, and batch systems offer the advantages of backup and processing error control.

On-line input allows the data entry operator to correct errors immediately. The edit checks capture inaccurate data at the time of entry and usually let the operator reenter the data. If, for any reason, the operator cannot rectify the errors, then, depending on the system, the rejected data may be stored in a file for correction in the future.

Since on-line data entry is done via interactive screens, the issues of screen formatting and human factors in screen design, discussed in Sections 7.6 and 7.7, apply to the on-line data entry process. In fact, screen displays are vehicles of both input data and output data.

7.9 CONVERSION OF A DATA FLOW DIAGRAM TO A SYSTEM FLOWCHART

We have seen in Chapter 5 that the end product of the system analysis phase contains a detailed layout of the proposed logical system. The data flow diagram is used as the graphic tool to describe the system. Since this tool is conceptual and totally independent of implementation considerations, it ideally suits a logical system description. At the system design stages, however, the logical system is converted into a physical system. Consequently, the data flow diagram is converted into a system flowchart, which is implementation bound. During the preliminary system design each data flow diagram should be carefully examined and then changed into an equivalent system flowchart. As a minimum, the level 1 data flow diagram must be converted into a system flowchart. This graphic then forms the basis of the physical system.

A data flow diagram has five main components:

a. Destination
b. Source

   c. Process

   d. Data flow

   e. Data store

The conversion of a data flow diagram into a system flowchart uses the following transformations:

32	a.	Source	→ Input
33	b.	Destination	→ Output
34	c.	Process	→ Program
35	d.	Data flow	→ Data element
36	e.	Data store	→ File

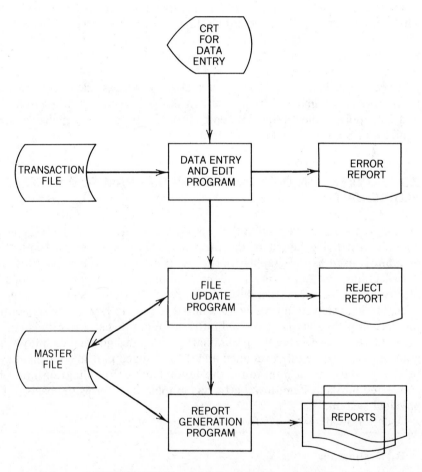

FIGURE 7-11   Example of a system flowchart.

Of course, these correspondences are not necessarily one to one; instead they are global in nature. This means that all the sources together should provide the complete input data, all the destinations together should give the output information, all the processes collectively should achieve the same goal as all the combined programs in the system flowchart, and all the data flows and data stores combined should account for all the fields and records in the system's data files.

Since the processes are the basic elements of a data flow diagram, the conversion starts by concentrating on the processes and identifying their functions. Then programs are conceived and categorized so that collectively the group of all processes becomes functionally equivalent to the group of all programs. Concomitantly, the data flows and the data stores used by the processes are examined. The data stores give rise to the data files used in the system flowchart. Also, the system team decides on the type of file processing involved, namely, sequential or random. As the files are designed, the team analyzes the data flows to create the data elements at the record level. (Recall that the data flows are essentially data elements at a group level.) During the detailed design phase the system team creates the record and file formats. Therefore, at the preliminary system design phase, it is enough to know which data stores are converted into which files.

Finally, the system team looks at the sources and the destinations. In most cases, the sources represent the point of entry of data into the system and as such are represented as data entry symbols, manual or on-line, on the system flowchart. The destinations, likewise, are represented as reports or display symbols in the system flowchart, because these are the system output. Thus, after the conversion is complete, the system team produces a system flowchart that gives an overview of the processing.

As an example, Figure 7-11 represents a system flowchart derived from the sample data flow diagram in Figure 3-2. The following table represents the cross-references used by the conversion process:

Data Flow Diagram (Fig. 3-2)	System Flowchart (Fig. 7-11)
Source: Data Entry Operator	Input: CRT for Data Entry
Destination: Recipients	Output: Reports
Processes 1 and 2	Data Entry/Edit Program
Process 3	File Update Program
Process 4	Report Generation Program
Data Store: Master File	Master File using random access
Data Store: Error File	Error Report from Data Entry/Edit Program
Data Store: Transaction File	Transaction File using random access

The data flows used in Figure 3-2 are implicitly contained in the Master File and the various reports.

7.10 SIMULATION OF A NEW SYSTEM VIA SYSTEM FLOWCHART

The system team should now simulate the flow of information through the new physical system. This is often called the *system validation process,* which is described below ([5], p. 34):

> The conceptual view of the system has now been converted into diagrams, flowcharts, narratives, and other supporting materials in order to prepare the design specifications.
>
> Any system design job is essentially a modeling job. Individual data entry and edit points, report dissemination processes, file updates, and flow of data through the system are captured and represented in the dataflow diagram, and the network topology. They are further quantified through the sizing analysis. This procedure replicates the real-life situation of the user's needs by stripping off some of their non-essential attributes and thereby making some simplifying assumptions. Consequently when the system is designed completely, it is necessary for the systems team to conduct a conceptual validation of the design. This consists of taking each user need separately and conducting a conceptual walk-through of the designed system to ensure that the need is indeed addressed properly by the system. This is similar to the validation process in a simulation model.

Let us now apply the system validation process by using Figure 7-11. The data entry operator enters data from a source document via the CRT. The data entry/edit program checks each field entered by applying the coded edit checks. If the data are error free, they are stored in the transaction file; otherwise, they are sent to the error report. Next, the file update program accepts the transaction records, compares them with the master records, and either updates the master records or sends unmatched records to the reject report. The updated master file is used by the report generation program to produce the necessary reports. Depending on the recommendations of the system team, either the data entry/edit program alone or both the data entry/edit program and the file update program will be run as on-line programs.

7.11 END PRODUCT OF PRELIMINARY SYSTEM DESIGN

Preliminary system design is the third phase of the system development process and marks the beginning of the system design activities. On completion of this phase, the system team should prepare a report containing the findings and recommendations from their study. Figure 7-12 provides a checklist of the items that should be included in the report. We now discuss these items.

> **1. Introduction.** Start with a thank you note for project acceptance, then provide a list of the precise capabilities (carryover from phase 2

1. Introduction
2. System flowcharts for the new system and its subsystems
3. List of all output reports with samples
4. List of all input documents with samples
5. Samples of screen displays (if applicable)

FIGURE 7-12   Checklist for findings report of phase 2.

report) that the new system will provide, and conclude with comments describing the transition from the analysis stage to the design stage, i.e., focus on the *how* issues instead of the *what* issues.

2. **System flowcharts.** Discuss why system flowcharts are used instead of the data flow diagrams of the earlier phases, describe the process of conversion of a data flow diagram into a system flowchart, and explain the objectives and the information flow of each system flowchart.

3. **Output reports.** Prepare a complete list of all the output reports to be produced by the system; for each report indicate the name, type (i.e., summary, exception, or detailed), frequency (i.e., daily, weekly, on request), and distribution (i.e., who receives it); finally provide samples of the report using a printer spacing chart.

4. **Input documents.** Prepare a complete list of all input documents used by the system, indicate if a document is a source document or a data entry sheet (in the latter case describe how data are gathered into the sheet from source documents), indicate the frequency of data entry using a given input document, and provide samples of the input documents.

5. **Screen displays.** If the system has an on-line component, include all screen displays and screen formats needed by the system; for each display describe how the user should interact with the system and describe the objective of the display.

7.12   SUMMARY

This chapter discussed the third phase of the system development process. It described the transition from the analysis to the design stage so that the emphasis is on *how* the capabilities listed during the analysis phase are going to be implemented. The logical system proposed at the end of the second phase will be converted to a physical system.

The system design activities are broken down into two phases: preliminary system design and detailed system design. Preliminary system design concentrates on detailed specifications of the input and output from the system and provides an overview of the processing aspect. The entire system is treated in a black box approach, with the processing component playing the role of the black box. The detailed system design phase will fully explore this black box. The difference between the two phases of system design is one of degree, not of kind.

The system output is of two types:

a. Printed report

b. Screen display

The printed report may be a summary report, a detailed report, or an exception report. Using a printer spacing chart, the system team prepares samples of all the reports. In addition, the team decides the frequency of generation of each report and its distribution. Figures 7-3 and 7-4 gave examples of report formats. The screen displays are more complex. They are required if there is an on-line component of the system. Due to the human interactions involved in the screen displays, the design must address issues such as

☐ Design consistency

☐ Dialogue design

☐ Performance criteria, especially response time

A wide variety of screen formatting software is currently available in the market. There is a definite trend toward using on-line systems. As a result, screen design has become an important activity.

The input documents should capture data as close to their origin as possible. Data may be entered directly from the source documents into the system or may first be transcribed onto a data entry sheet and then entered into the system. If there is on-line data entry, an appropriate edit program should check and verify data during the data entry process.

The final step in the preliminary system design is to prepare an overview of the processing component. The system flowchart is used as the graphic tool, because the new system is now a physical system that is implementation bound. Consequently, the data flow diagrams describing the proposed logical system are now converted into the corresponding system flowcharts. This conversion process uses the following transformations:

$$
\begin{aligned}
\text{Source} &\rightarrow \text{Input} \\
\text{Destination} &\rightarrow \text{Output} \\
\text{Process} &\rightarrow \text{Program} \\
\text{Data flow} &\rightarrow \text{Data element} \\
\text{Data store} &\rightarrow \text{File}
\end{aligned}
$$

The basic principle is to guarantee that, collectively, all the processes combined will accomplish the same functions as all the programs together, all the data stores together must contain the same total set of data as all the files combined, and so on. Finally, the system team should simulate the flow of information through the system by using the system flowcharts. This is often called the system validation.

The chapter closed with a brief description of the items that should be included in the final report summarizing the findings and recommendations of the preliminary system design phase.

7.13 KEY WORDS

The following key words are used in this chapter:

black box approach
data entry sheet
design consistency
detailed report
dialogue design
exception report
functional level of screen design
Hansen's user engineering principles
logical system
Mooers' law
performance criteria
physical system

printed report
printer spacing chart
procedural level of screen design
response time
screen display
screen formatting
source documents
summary report
syntactical level of screen design
system flowchart
system validation process

REFERENCES

1. W. O. Galitz, *Handbook of Screen Format Design*, QED Information Sciences, Wellesley, MA, 1981.
2. F. W. Lancaster and E. G. Fayen, *Information Retrieval On-Line*, Melville Publishing Company, Los Angeles, 1973.
3. Henry C. Lucas, Jr., *Analysis, Design and Implementation of Information Systems*, McGraw Hill, New York, 1985.
4. James Martin, *Design of Man—Computer Dialogues*, Prentice-Hall, Englewood Cliffs, NJ, 1973.
5. Sitansu S. Mittra, "Information Systems Analysis and Design," *Journal of Systems Management*, April 1983, pp. 30–34.
6. James C. Wetherbe, Jr., *System Analysis for Computer-Based Information System*, West Publishing Company, St. Paul, MN, 1979.

REVIEW QUESTIONS

1. How does system design differ from system analysis? Why do we need two stages of system design?
2. What is meant by a black box approach to system design?
3. Explain the three different types of printed reports. Give an example of each.
4. Why is the human factor a major consideration in screen design? Describe the three levels of user psychology involved in screen design.
5. Comment on Mooers' law and give examples.
6. What is meant by the design consistency of screens? Give examples.
7. Discuss the basic principles behind dialogue design.
8. Define response time, and explain the rule of thumb of response time requirements.
9. What is an efficiency ratio? Is it reliable? Give reasons for your answer.
10. What are the main objectives of screen formatting software?
11. Explain the difference between a source document and a data entry sheet.
12. Explain the conversion process from a data flow diagram to a system flowchart.
13. What is meant by saying that the correspondence between the components of a data flow diagram and those of a system flowchart is not necessarily one-to-one but only global in nature?
14. What is system validation? Why is it needed?
15. What should be the components of the end product of the preliminary system design?

8

Preliminary System Design: Case Studies

8.1 INTRODUCTION

In this chapter we shall discuss the preliminary system design phases of the two case studies introduced in Chapter 4 and further explored in Chapter 6. Using the checklist given in Figure 7-12, we shall prepare the items to be included in the end product of each case study.

Toy World, Inc., requires an order processing system with linkages to inventory control and sales processing. Massachusetts Educational Foundation needs a financial reporting system. The preliminary system design reports for both systems are now described.

8.2 PRELIMINARY SYSTEM DESIGN REPORT FOR TOY WORLD

8.2.1 Letter of Appreciation

Figure 8-1 shows the letter sent to Mr. Dennis Friend, president of Toy World, from HPC expressing their appreciation for the acceptance of HPC's proposal. In addition, the letter reinforces the capabilities to be provided by the new system.

8.2.2 System Flowcharts

Figure 4-3 represents the level 1 data flow diagram for the new system. Four of the six processes have been exploded at level 2 in Figures 6-5 through 6-8. HPC now converts the six processes into six programs for the physical system. Figures 8-2 through 8-7 represent the system flowcharts for these six subsystems. A brief description of each process follows:

Objectives of the System Flowcharts

Process 1. To illustrate the order entry process

Process 2. To illustrate the processing of inventory change transactions

Process 3. To illustrate the generation of the sales report

Process 4. To illustrate the processing of customer requested changes

Process 5. To illustrate the order selection process and generation of the picking list and detailed invoices

Process 6. To illustrate the generation of the inventory status report

8.2.3 Output Reports

HPC recommends that four reports be generated on a regular basis from the order processing system. In addition, there may be on-demand reports or queries of an ad hoc nature. The following list gives the names, frequencies, and recipients of the four prescheduled reports:

Name	Frequency	Recipient
Picking List	Weekly	Warehouse
Detailed Invoice	Weekly	Warehouse
Sales Statistics Report	Biweekly	Management
Inventory Status Report	Biweekly	Management, Warehouse

Figures 8-8 through 8-11 give samples of these reports.

8.2.4 Input Documents

In order to generate the above reports and maintain the system, it is necessary to collect data via these input documents:

1. Order Form
2. Customer Change Transaction
3. Management Change Transaction

Figures 8-12 through 8-14 give examples of these input documents.

$ H P C

October 22, 1983

Mr. Dennis Friend, President
Toy World, Inc.
Harrisburg, PA.

Dear Mr. Friend :

HPC is especially pleased with your acceptance of our approach
to implement the proposed system in your organization.

The HPC system will not only provide your organization with
order-filling information utilized by your warehouse crew; it
will also update your inventory records based upon the amount
of actual shipments of your product to customers; moreover,
the system will retain summary data for present and future
invoicing applications. The Inventory Control System will
maintain all the inventory at a proper level not only to meet
customer requirements, but, in addition, ensuring that company
has a minimum invested in its inventory.

The HPC system will provide the Toy World, Inc.   with the
following major capabilities:

    1) Customer Analysis
    2) Product Analysis
    3) Sales Analysis
    4) Sales/Market Forcasting (trends)
    5) Sales District Analysis
    6) Generation of Inventory Status Reports
    7) Improved Product Control
    8) Improved Production Operation
    9) On-Line Order Entry
    10) On-Line Customer Information Displays
    11) Programmed Substitution Rules

In effect, the HPC system will assist the Toy World, Inc.
in its efforts to increase the organization's sales and profits
and with this understanding we at HPC, are anticipating our
next meeting with you.

Sincerely yours

High Price Consultants

FIGURE 8-1 Letter of appreciation from HPC.

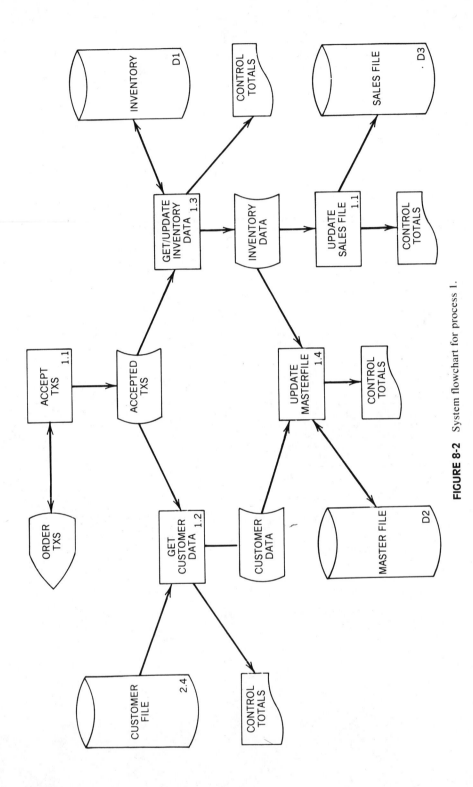

FIGURE 8-2 System flowchart for process 1.

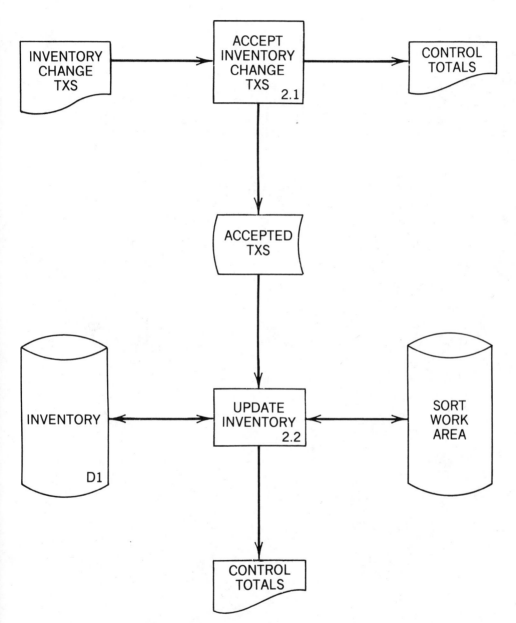

FIGURE 8-3 System flowchart for process 2.

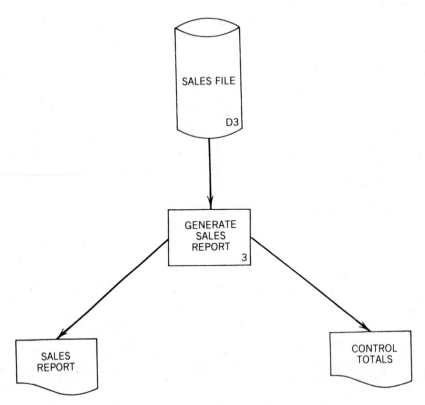

FIGURE 8-4 System flowchart for process 3.

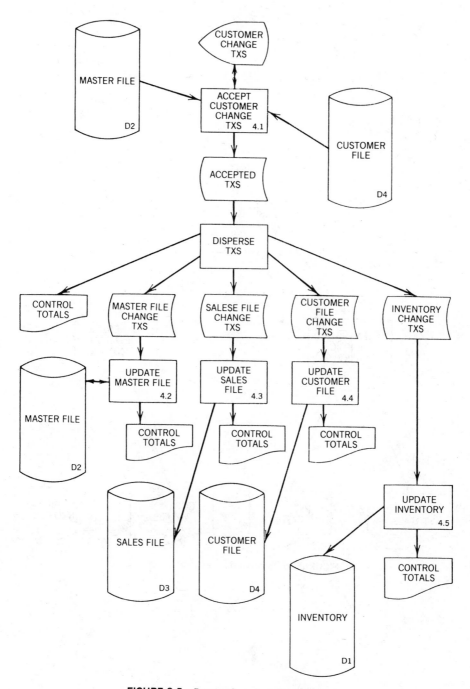

FIGURE 8-5 System flowchart for process 4.

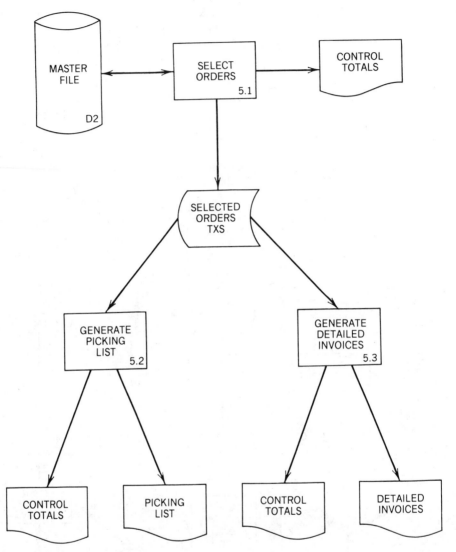

FIGURE 8-6 System flowchart for process 5.

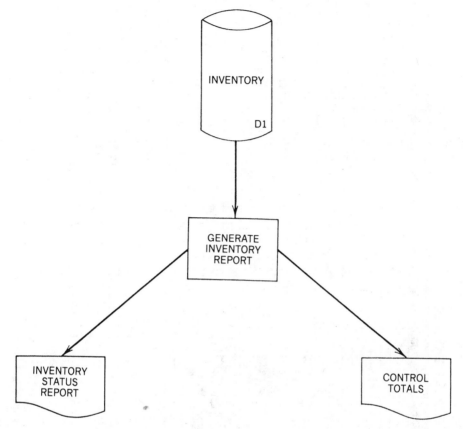

FIGURE 8-7 System flowchart for process 6.

REPORT LAYOUT

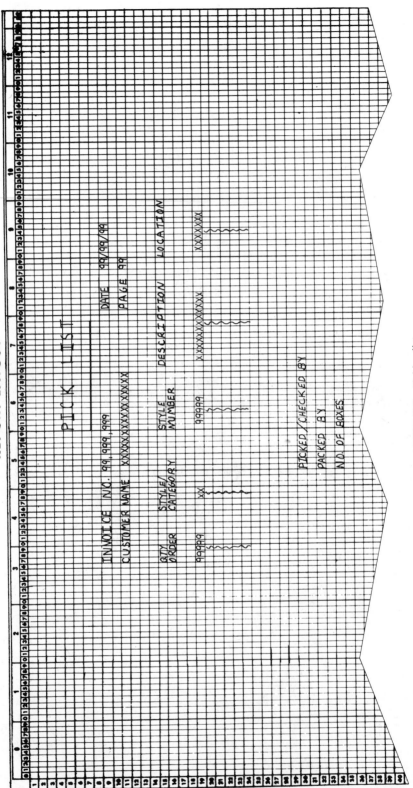

FIGURE 8-8 Picking list.

REPORT LAYOUT

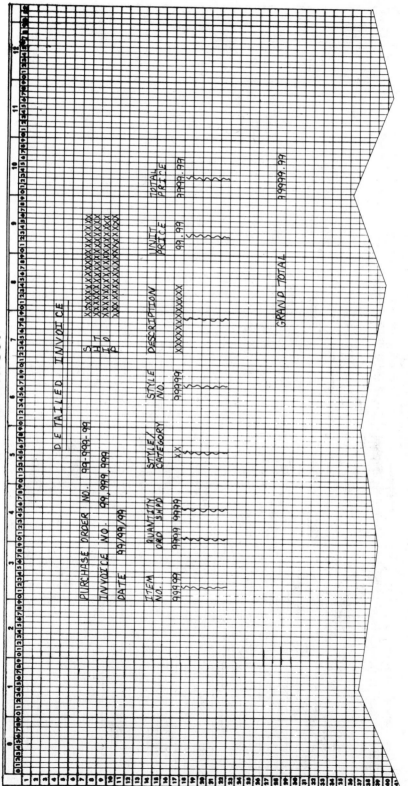

FIGURE 8-9 Detailed invoice.

REPORT LAYOUT

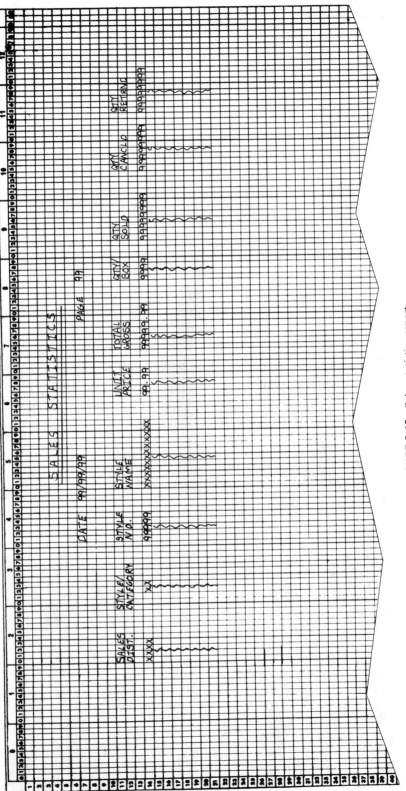

FIGURE 8-10 Sales statistics report.

REPORT LAYOUT

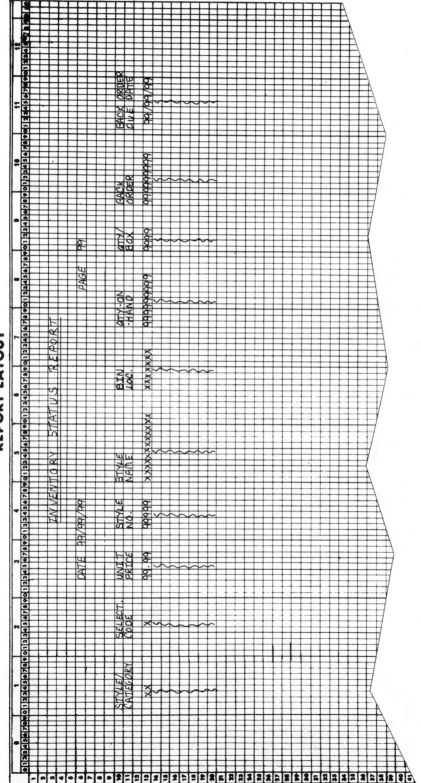

FIGURE 8-11 Inventory status report.

TOY WORLD, INC.

HARRISBURG

PENNSYLVANIA

This No. must appear on all invoices, packing slips, and packages.

| PURCHASE ORDER NO. | 99-999-99 |
| CHANGE ORDER NO. | |

V E N D O R		S H I P T O	
XXX XXXXXXXX		XXXXXX XXXXXXXXXXXXX	
999 XXXXX XX		99 XXXX XX	
XXXXXXXX XXXX		XXXXXXXX,XXX 99999	

Confirming Order ☐ Yes ☐ No

Date	Requisition No.	Ship Via	Ppd ☐	F.O.B	Terms	Basic Agreement No	Delivery Date
99/99/99	9999		Coll ☐				99/99/99

THIS ORDER IS SUBJECT TO THE CONDITIONS ON THE FACE AND REVERSE SIDE

Item	Quantity	Purch	Stock Number—Description	Unit Price	Amount
99	99		999-999 XXXXX XX	$$$.$$	$$$$.$$
9	99		999-99 XXXX XXX	$$$.$$	$$$$.$$
			XXXXXXXXXXXX XXXXX		
			XXXXX		$$$$.$$

INVOICE IN DUPLICATE
TO ACCOUNT PAY
DEPARTMENT

SPECIAL INSTRUCTIONS:	Sales Tax	Tax Exempt No 04-2226590-Mass	FOR FURTHER INFORMATION CONTACT:
	☐ Tax ☐ No Tax	Other_____	

ACCEPTANCE

Your order as given above is hereby acknowledged and accepted

Signed_____

For_____
(Company)

FOR PURCHASING USE ONLY	
Buyer Signature	Date
Approval Signature	Date

PURCHASING COPY

FIGURE 8-12   Order form.

8.2.5 Screen Displays

HPC recommends that the following screen displays be used in the on-line part of the system:

1. User Login
2. Sales Order Entry
3. New Order

TOY WORLD, INC.

CUSTOMER CHANGE TRANSACTION

DATE: _____   ____LEC/LAB   ____ SEMINARS
REFERRED BY: _____   ____ AV/SPI   ____OTHER
CALL TAKEN BY: _____   ____DECPRESS   ____NON-E/S

CUSTOMER	SALES
(GSA_____)	

NAME _____   REP_____
TITLE_____   OFFICE _____
COMPANY_____   DTN# _____
ADDRESS _____   TEL.# _____
_____
                          (Zip Code)
_____

MARKETING REP: _____   LOC _____ M/S_____
TEL.#_____   (DTN)_____

ADDITIONAL INFORMATION: _____
_____
_____
_____
_____
_____
_____
_____
_____

ACTION TAKEN: _____
_____
_____
_____
_____
_____
_____
_____
_____

TRANSFERRED/REFERRED TO: _____   DATE: _____
FOLLOW-UP DATE: _____
CLOSED DATE: _____

FIGURE 8-13 Customer change transaction.

4. Customer Service Menu
5. Change Customer Data Menu
6. Change Customer Name
7. Change Customer Address
8. Change Customer Phone Number
9. Cancel Order
10. Alter Order

Figures 8-15 through 8-24 give examples of these screen displays.

TOY WORLD, INC.

MANAGEMENT CHANGE TRANSACTION

DATE: _____	____ LEC/LAB ____ SEMINARS
REFERRED BY: _____	____ AV/SPI ____ OTHER
CALL TAKEN BY: _____	____ DECPRESS ____ NON-E/S

MANAGEMENT (GSA_____)	**SALES**
NAME _____	REP_____
TITLE _____	OFFICE _____
COMPANY_____	DTN# _____
ADDRESS _____	TEL.# _____
_____ (Zip Code)	
_____	

MARKETING REP: _____   LOC _____ M/S _____
TEL.#_____   (DTN)_____

ADDITIONAL INFORMATION: _____
_____
_____
_____
_____
_____
_____
_____
_____

ACTION TAKEN: _____
_____
_____
_____
_____
_____
_____
_____
_____

TRANSFERRED/REFERRED TO: _____   DATE: _____
FOLLOW-UP DATE: _____
CLOSED DATE: _____

FIGURE 8-14   Management change transaction.

8.3   PRELIMINARY SYSTEM DESIGN FOR MASSACHUSETTS EDUCATIONAL FOUNDATION

After getting approval on their system analysis report for the Massachusetts Educational Foundation (MEF), HPC completed the preliminary system design phase for the Financial Reporting System. The end product of their study is described here.

8.3.1 System Flowchart

Figure 4-5 gave the level 1 data flow diagram for the system. Figure 8-25 represents the system flowchart corresponding to Figure 4-5. We notice the following correspondence between the entities of the data flow diagram and those of the system flowchart:

Data Flow Diagram (Fig. 4-5)	System Flowchart (Fig. 8-25)
Source: USER	Input: New forms or previously rejected forms
Destination: USER	Output: Annual Report
Process 1	Data Entry/Edit Program
Process 2	Update Forms Database Program
Process 3	Validation Program
Process 4	Ad hoc Query Program
Process 5	Annual Report Generation Program
Data Store: D1	Edited Forms File using random access and Form Database
Data Store: D2	Error Report
Data Store: D3	Form Database
Data Store: D4	Nonvalidated Forms Report

Thus, collectively the high-level system flowchart conveys the same information as the level 1 data flow diagram. Since the system provides substantial on-line capabilities (see Section 6.11), the executive module appears in Figure 8-25 although it has no counterpart in Figure 4-5. When an on-line user logs in, he or she gains access to the system via the executive module (see Section 4.11).

Finally, Figure 8-26 represents the system flowchart corresponding to the two data flow diagrams given in Figures 6-16 and 6-17.

8.3.2 Output Reports

The primary output report is the Annual Report, which is generated once a year. It consists of over 200 statistical tables dealing with financial data. Other prescheduled output reports are listed below:

Name	Frequency	Recipient
Error Report	Daily	Data analysts
Nonvalidated Forms List	Daily	Validation clerks

Figures 8-27 and 8-28 show examples of these reports.

8.3.3 Input Documents

MEF uses a wide variety of forms to gather the necessary data from its grantees. As samples, the three forms MEF-04, MEF-10, and MEF-25, which are mentioned in Figure 8-28, are shown in Figures 8-29, 8-30, and 8-31, respectively.

8.3.4 Screen Displays

The detailed interactions of the screen displays have been discussed in Sections 4.11 and 4.12. As samples, we include two screens in Figures 8-32 and 8-33.

8.4 SUMMARY

This chapter completed the preliminary system designs for the two case studies, Toy World and Massachusetts Educational Foundation, by applying the theory developed in Chapter 7 to them.

For Toy World, Inc., the chapter contained the following items:

a. Six system flowcharts showing the implementations of the six processes discussed in Chapters 4 and 6
b. Four output reports
c. Three source documents providing input data
d. Ten screen displays to handle the on-line part of the system

Figures 8-2 through 8-24 represent the above items.

For Massachusetts Educational Foundation, the chapter contained the following items:

a. Two system flowcharts, one corresponding to a macro level data flow diagram (Fig. 4-5) and another corresponding to two micro level data flow diagrams (Figs. 6-16 and 6-17)
b. Two output reports
c. Three sample forms as input documents
d. Two screen displays showing two versions of the main menu

8.5 KEY WORDS

The following key words are used in this chapter:

data flow diagram	screen display
input document	system flowchart
preliminary system design	system validation
process	

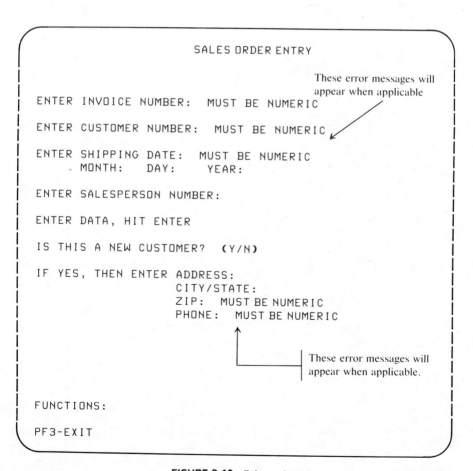

```
ENTER PASSWORD:

ENTER EMPLOYEE NUMBER:




ENTER DATA, HIT ENTER


INVALID PASSWORD

INVALID EMPLOYEE NUMBER
```
These error messages will
appear when applicable.

FIGURE 8-15   User login.

```
                    SALES ORDER ENTRY
```
These error messages will
appear when applicable
```
ENTER INVOICE NUMBER:   MUST BE NUMERIC

ENTER CUSTOMER NUMBER:   MUST BE NUMERIC

ENTER SHIPPING DATE:   MUST BE NUMERIC
      MONTH:    DAY:     YEAR:

ENTER SALESPERSON NUMBER:

ENTER DATA, HIT ENTER

IS THIS A NEW CUSTOMER?  (Y/N)

IF YES, THEN ENTER ADDRESS:
                CITY/STATE:
                ZIP:  MUST BE NUMERIC
                PHONE:  MUST BE NUMERIC
```
These error messages will
appear when applicable.
```

FUNCTIONS:

PF3-EXIT
```

FIGURE 8-16   Sales order entry.

```
                          NEW ORDER


     CUSTOMER NAME   XXXXXXXXXXXXXXXXXXXX

     CUSTOMER NUMBER  999999

     INVOICE NUMBER  99999999

     ENTER STYLE CATEGORY:

     ENTER STYLE NUMBER:

     ENTER QUANTITY:

     ENTER DATA, HIT ENTER


     ARE THERE MORE ITEMS TO BE ORDERED ON

     THIS INVOICE (Y/N):  HIT ENTER


     INVALID REPLY, ENTER Y OR N

     STYLE NUMBER MUST BE NUMERIC:          This error message will
                                            appear when applicable.


     FUNCTIONS:

     PF1-HELP PF2-EXIT
```

FIGURE 8-17 New order.

```
                    CUSTOMER SERVICE MENU


     ENTER CUSTOMER NUMBER:

                 OR NAME:

     SELECTION CATEGORY

     (1) CHANGE CUSTOMER DATA MENU
     (2) CANCEL ORDER
     (3) ALTER ORDER
     (4) DISPLAY CUSTOMER DATA
     (5) EXIT

     ENTER SELECTION NUMBER:  HIT ENTER


     INVALID SELECTION NUMBER

     CUSTOMER NUMBER NOT IN FILE

     CUSTOMER NAME NOT IN FILE
                                       These error messages will
                                       appear when applicable.




     FUNCTIONS:

     PF1-HELP PF2-DELETE INPUT
```

FIGURE 8-18 Customer service menu.

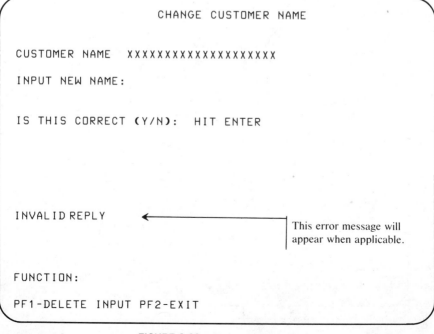

```
┌─────────────────────────────────────────────────────────────┐
│                CHANGE CUSTOMER DATA MENU                      │
│                                                               │
│   SELECTION CATEGORY                                          │
│                                                               │
│   (1) CHANGE CUSTOMER NAME                                    │
│                                                               │
│   (2) CHANGE CUSTOMER ADDRESS                                 │
│                                                               │
│   (3) CHANGE CUSTOMER PHONE                                   │
│                                                               │
│   (4) EXIT                                                    │
│                                                               │
│                                                               │
│   ENTER SELECTION NUMBER, HIT ENTER                           │
│                                                               │
│   INVALID SELECTION NUMBER    ◄─────                          │
│                                    This error message will    │
│                                    appear when applicable.    │
│                                                               │
│   FUNCTION:                                                   │
│   PF1-HELP                                                    │
└─────────────────────────────────────────────────────────────┘
```

FIGURE 8-19   Change customer data menu.

```
┌─────────────────────────────────────────────────────────────┐
│                  CHANGE CUSTOMER NAME                         │
│                                                               │
│   CUSTOMER NAME   XXXXXXXXXXXXXXXXXXXX                         │
│   INPUT NEW NAME:                                             │
│                                                               │
│   IS THIS CORRECT (Y/N):   HIT ENTER                          │
│                                                               │
│                                                               │
│                                                               │
│   INVALID REPLY    ◄─────                                     │
│                         This error message will               │
│                         appear when applicable.               │
│                                                               │
│   FUNCTION:                                                   │
│   PF1-DELETE INPUT PF2-EXIT                                   │
└─────────────────────────────────────────────────────────────┘
```

FIGURE 8-20   Change customer name.

```
                    CHANGE CUSTOMER ADDRESS


CUSTOMER ADDRESS    XXXXXXXXXXXXXXXXXXXXX

                    XXXXXXXXXXXXXXXXXX

                                99999

INPUT NEW ADDRESS:

          STREET:

       CITY/STATE:

             ZIP:

    IS THIS CORRECT:   HIT ENTER

INVALID REPLY, ENTER Y OR N

ZIP CODE MUST BE NUMERIC

STREET MUST BE REENTERED

CITY/STATE MUST BE REENTERED

ZIP CODE MUST BE REENTERED
                                    These error messages will
                                    appear when applicable.


PF1-DELETE INPUT PF2-EXIT
```

FIGURE 8-21 Change customer address.

```
                    CHANGE CUSTOMER PHONE NUMBER


CUSTOMER PHONE NUMBER   XXX-XXX-XXXX

INPUT NEW PHONE NUMBER:

IS THIS CORRECT (Y/N):   HIT ENTER




INVALID REPLY
PHONE NUMBER MUST BE NUMERIC
                                    These error messages will
                                    appear when applicable.



FUNCTIONS

PF1-DELETE INPUT PF2-EXIT
```

FIGURE 8-22 Change customer phone number.

```
                        CANCEL ORDER


 ENTER INVOICE NUMBER:                 HIT ENTER

 CUSTOMER NUMBER 999999

 CUSTOMER NAME   XXXXXXXXXXXXXXXXXXXX

 DATE ORDER RECEIVED   XX/XX/XX

 DATE SHIPMENT REQUESTED   XX/XX/XX

 QUANTITY ORDERED   99999



 INVOICE NUMBER NOT IN FILE

 INVOICE NUMBER MUST BE NUMERIC
                                        These error messages will
                                        appear when applicable.


 FUNCTIONS:

 PF1-HELP PF2-EXIT
```

FIGURE 8-23 Cancel order.

```
                        ALTER ORDER


 ENTER INVOICE NUMBER:                 HIT ENTER

 CUSTOMER NUMBER:   999999

 CUSTOMER NAME:   XXXXXXXXXXXXXXXXXXXX

 ENTER STYLE NUMBER:   MUST BE NUMERIC

 ADD OR DELETE (A,D):   A OR D ONLY

 QUANTITY OF BOXES:   MUST BE NUMERIC

 IS THE INFORMATION CORRECT (Y/N)   Y OR N ONLY

 MORE CHANGES (Y/N):   Y OR N ONLY


 INVOICE NUMBER NOT IN FILE

 INVOICE NUMBER MUST BE NUMERIC

 STYLE NUMBER NOT IN FILE
                                        These error messages will
                                        appear when applicable.


 FUNCTIONS:

 PF1-HELP PF2-EXIT
```

FIGURE 8-24 Alter order.

REVIEW QUESTIONS

Complete the preliminary system flowcharts corresponding to the design study for the Massachusetts Educational Foundation following these guidelines:

1. Prepare system flowcharts corresponding to the micro level data flow diagrams (see Chapter 6).
2. Generate additional output reports.
3. Design a screen to implement the part of the Executive Module (see Section 6.11) that precedes the two versions of the main menu given in Figures 8-32 and 8-33. What function does this part implement?
4. Design two additional screens to implement the data entry/edit/display option in Figure 8-32.
5. Design three additional screens to implement the database administration option in Figure 8-33. (Note: You will need good knowledge of database systems.)
6. Apply the system validation process by simulating the flow of information through the system. For example, you can start with the data entry from the form MEF-04 into the system and end up with generating the report given in Figure 8-28. Use the system flowchart in Figure 8-25 and appropriate screens to trace the information flow through the system. See Section 7.10 for a description of the system validation process.

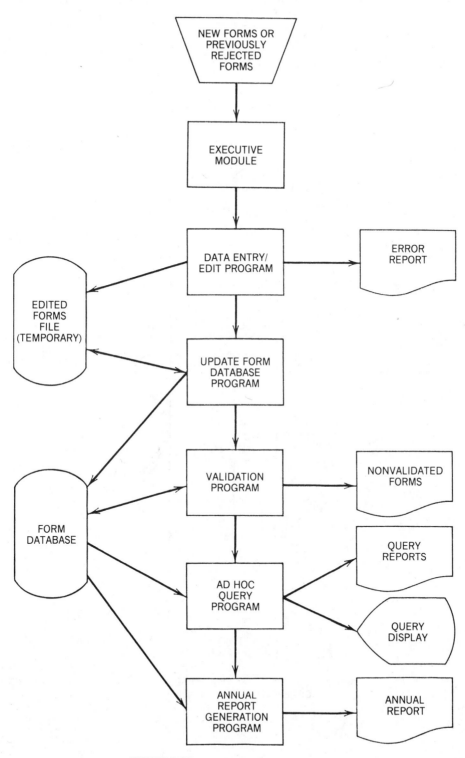

FIGURE 8-25 System flowchart for MEF.

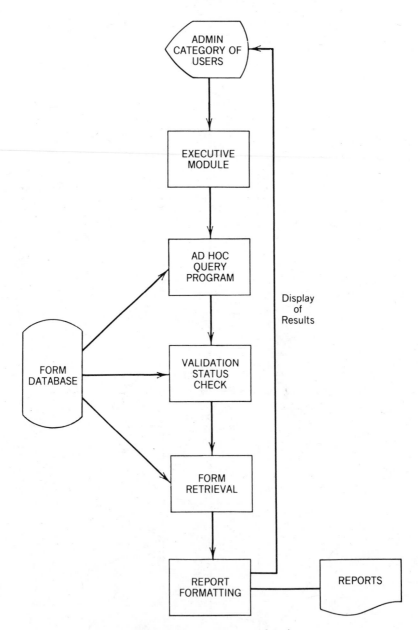

FIGURE 8-26   On-line query function.

MASSACHUSETTS EDUCATIONAL FOUNDATION

ERROR REPORT FOR FORM XXXXX

DATE   XX/XX/XX

LINE No.	COLUMN No.	ERROR CODE	ERROR DESCRIPTION
1	a	12	Data must be numeric
4	e	04	Data out of range
9	c	19	Invalid date
.	.	.	-----
.	.	.	-----
.	.	.	-----
.	.	.	-----

FIGURE 8-27 Sample error report.

MASSACHUSETTS EDUCATIONAL FOUNDATION

NON_VALIDATED FORMS LIST

DATE   XX/XX/XX   PAGE X

FORM No.	VALIDATION CODE AND LOCATION OF VIOLATION	REASON
MEF-25	7 - Line 4, column c	Value exceeds 30% of Line 4, column d
MEF-04	2 - Line 4, Column d and Line 5, Column d }	Values do not add up to 100%
MEF-04	3 - MEF-04 (Line 5,Col. d)	Values are not
MEF-10	and MEF-10 (Line4, Col. c) }	equal
.	.	.
.	.	.

FIGURE 8-28 Sample nonvalidated forms lists.

MASSACHUSETTS EDUCATIONAL FOUNDATION

FORM MEF-04

FUNDING SOURCES

GRANTEE ID _____   FISCAL YEAR _____   CATEGORY _____

LINE   COLUMN

                          (a)   (b)   (c)   (d)

                                                        (= (a) + (b) + (c))

1. Federal Grant   _____   _____   _____   _____

2. State Grant   _____   _____   _____   _____

3. City/Town Grant   _____   _____   _____   _____

4. Total Public Grant
   (= Sum of lines 1,2,3) _____   _____   _____   _____

5. Total Private Grant   _____   _____   _____   _____

FIGURE 8-29 Sample form MEF-04.

MASSACHUSETTS EDUCATIONAL FOUNDATION

FORM   MEF-10

PRIVATE FUNDING BREAKDOWN

GRANTEE _____   FISCAL YEAR _____   CATEGORY _____

LINE   COLUMN

                                        (a)   (b)   (c)

                                                              (= (a) + (b))

1. Funding Agencies

   within Grantee State   _____   _____   _____

2. Funding Agencies outside

   Grantee State,

   but within USA   _____   _____   _____

3. Non-USA Funding Agencies   _____   _____   _____

4. Total Private Funding

   (= Sum of lines 1,2,3)   _____   _____   _____

FIGURE 8-30 Sample form MEF-10.

MASSACHUSETTS EDUCATIONAL FOUNDATION

FORM   MEF-25

GRANTEE EMPLOYEE EXPENSE

GRANTEE ID _____   FISCAL YEAR _____   CATEGORY _____

LINE   COLUMN

                  (a)   (b)   (c)   (d)

                Clerical   Technical   Supervisory   Total

                                                    $(= (a) + (b) + (c))$

1. Salary   _____   _____   _____   _____

2. Benefits   _____   _____   _____   _____

3. Other

Compensations   _____   _____   _____   _____

4. Total (= Sum

   of Lines 1,2,3) _____   _____   _____   _____

FIGURE 8-31   Sample form MEF-25.

MASSACHUSETTS EDUCATIONAL FOUNDATION

MAIN   MENU

(USER   VERSION)

The following options are available :

1.   Data Entry/Edit/Display

2.   Data Validation

3.   Help

4.   Exit

Select an option and type the corresponding number. Then press

RETURN.

FIGURE 8-32   Main menu (USER version).

MASSACHUSETTS EDUCATIONAL FOUNDATION

MAIN   MENU

(ADMIN   VERSION)

The following options are available :

1.   Data Entry/Edit/Display

2.   Data Validation

3.   Database Administration

4.   Help

5.   Exit

Select an option and type the corresponding number. Then
press   RETURN.

FIGURE 8-33   Main menu (ADMIN version).

9

Detailed
System Design:
Theory

9.1 ISSUES ADDRESSED UNDER DETAILED SYSTEM DESIGN

The detailed system design carries out the system design process started in
Chapter 7 to its fullest detail. The difference between this phase and the
preliminary design phase is one of degree only. The situation is similar to
the distinction between the first and the second phases of system development
(see Chapters 3 and 5). A prerequisite for the detailed system design phase
is the identification by the user of one specific solution out of two or more
submitted by the systems team.

In the detailed design phase, the team works with the solution selected
by the user and prepares detailed design specifications showing how the
system can be implemented. The specifications are analogous to an engineer's
blueprints for constructing a building. Given these detailed specifications,
the programmers can write their programs and the team leader can prepare
precise cost estimates and implementation schedule. Moreover, the user can
see an image of the proposed system.

Detailed system design concentrates on the file structure by starting with
the conversion of data flows into data elements at the record level, specifying
the file access and file organization methods (i.e., sequential, indexed sequential, or random), and estimating auxiliary storage needed for the files.
It addresses the issue of using database systems as an alternative to file
systems. It explores a quantification methodology to estimate auxiliary storage. Finally, it describes rudimentary data communication techniques needed

to design a network topology that may be needed for a distributed or de-centralized system. Consequently, the following topics are discussed in this chapter:

1. Data elements with formats
2. Record and file structure
3. File organization—sequential, indexed sequential, or direct
4. Schema structure, if a database is involved
5. Space estimate for files
6. Auxiliary storage estimate
7. Design of data communication network and estimate of traffic volume, if applicable
8. Equipment specification with a configuration chart
9. Personnel selection

9.2 DATA ELEMENTS AT RECORD LEVEL

Data stores in a data flow diagram are converted to data files (see Section 7.9). It is necessary to convert the individual data flows within a given data store into data elements at the record level within a file. In general, a single data flow contains enough data to produce multiple data elements. Thus, the entry for a data flow in the data dictionary must be analyzed and subdivided into several data elements. For example, the data dictionary may contain a data flow called CUSTOMER DATA. At a record level, this must be further broken down into multiple data elements such as

```
CUSTOMER NAME
CUSTOMER ADDRESS
PHONE NUMBER
```

In other words, the correspondence between a data flow and its underlying data elements is one to many. The system designer must prepare a comprehensive list of data elements from the set of data flow entries in the data dictionary. This can be done as follows:

a. Select a data file whose data elements are to be designed.
b. Determine the data stores that are used to create the file.
c. Identify in the data dictionary all the data flows that belong to the data stores.
d. Expand each data flow until you reach the data element level. This step is critical and somewhat subjective.

e. Decide on the name, format, and description of each data element. Ensure that each data element has a *unique* name.

f. Combine all the relevant data elements that constitute the record structure.

9.3   RECORD AND FILE FORMAT

The design of files is the three data layout process, the first two steps being the design data elements and the design of the record structure. A record consists of a group of data elements pertaining to a specific area; for example, a payroll record contains data elements such as employee name, Social Security number, job classification, and salary level. A file consists of the set of all records related to a specific application; for example, a payroll file consists of the payroll records of all the employees in a company.

A record may have a fixed or variable length. In a fixed-length record, we know in advance the size of every data element and the total number of the data elements in the record. In a variable-length record, the size of each data element is known beforehand but not the total number of data elements in the record. Typically, a variable-length record has two segments: a fixed-length segment, which is similar to a fixed-length record, and a variable-length segment, which consists of repeated occurrences of some predetermined set of data elements with the number of occurrences subject to a preestablished maximum value.

Figure 9-1 gives an example of a fixed-length record in a payroll file. It is necessary to specify the name, description, and format of each data element in the record structure.

The need for a variable-length record arises when some of the data elements in the record may be repeated. For example, consider a system that keeps track of patient visits to a doctor's office. This system needs a patient history file with records that contain patient data such as name, address, insurance, diagnosis, test results, and drugs prescribed. The data elements such as patient name, address, and so on, constitute the fixed segment of the record, since they appear only once in each record. The diagnosis, drugs, test results, and so on, appear once for each visit of the patient. Accordingly, they belong to the variable segment of the record. Arbitrarily, we may decide that normally a patient does not visit more than six times a year. Then we decide that the variable segment can be repeated at most six times in a record and that the record is reinitialized at the beginning of each year, with the old record being archived at that time.

Figure 9-2 gives an example of a variable-length record. The first four data elements, NAME, ADDRESS, PHONE, and ACCTNO, belong to the fixed segment, whereas the last four data elements, SDATE, SCODE, DRUG, and CHRG, constitute the variable segment that can appear up to six times within the record.

Name	Description/Comments	Size and Format
NAME	Last name, first name, middle initial	20 characters; alphanumeric
SSNO	Social Security number	11 characters; alphanumeric
JOB-CODE	Numeric code for job classification	4 characters; numeric
HRLY-RATE	Hourly rate of compensation	5 characters; numeric (last two digits are cents)
HRS-wrk	Number of hours worked, regular and overtime	4 characters; numeric (last two digits are after decimal)
EXEMPTION	Number of exemptions claimed	2 characters; numeric
FED-TAX	Federal tax withheld	6 characters; numeric (last two digits are cents)
ST-TAX	State tax withheld	6 characters; numeric (last two digits are cents)
FICA	Social Security tax withheld	6 characters; numeric
FILLER	Reserved for future expansion	16 characters; alphanumeric
TOTAL		80 characters

FIGURE 9-1 Example of fixed-length record.

Name	Description/Comments	Size and Format
NAME	Last name, first name, middle initial	20 characters; alphanumeric
ADDRESS	Street and number, city, state, zip code	60 characters; alphanumeric
PHONE	Area code and number	10 characters; numeric
ACCTNO	Account number (unique for each patient)	5 characters; numeric
*SDATE	Date visiting doctor	6 characters; numeric in MMDDYY format
*SCODE	Code describing the diagnosis	8 characters; alphanumeric
*DRUG	Code describing drug administerd; blank if no drug given	8 characters; alphanumeric
*CHRG	Total charge for the visit	5 characters; numeric (last two digits are cents)

*The data elements can be be repeated up to six times.

FIGURE 9-2 Example of variable-length record.

9.4   FILE ACCESS AND FILE ORGANIZATION

After the system team prepares a physical layout of each file, it has to determine the file access and the file organization method. There are two file access methods: sequential and direct, or random. In *sequential access,* each record is accessed sequentially; for example, in order to access the Nth record, it is necessary to access the first $N - 1$ records. In *direct access,* the system can access the Nth record directly without referring to the first $N - 1$ records.

There are three separate file organization methods: sequential, random, and indexed sequential. As a *sequentially organized* file is created, records are added one after another in a chainlike fashion. Later, when the records are processed, they are read in the same fixed sequence. Records can be stored in order or sorted and stored by some key. The basic idea is that a sequential file must be processed in a fixed order. Hence the speed with which a given record can be accessed depends upon its physical location within the file. A sequentially organized file can only be accessed sequentially.

In a *randomly organized file,* each logical record is assigned a key that corresponds in some way to the physical location of the data on disk. Given this key, it is possible to store or retrieve a record without regard for its position on the file; for example, record N can be accessed just as quickly and just as easily as any other record. However, the contents of such a file cannot be retrieved in any meaningful sequential order. If it is necessary to do so, the records have to be copied into a new file, sorted there, and then accessed sequentially.

A third organization method is an *indexed sequential file.* As the data are stored, an index is created linking each record's logical key to its physical address on disk; the index is then stored independently. Later, when the data must be retrieved, the index is read into main memory and searched by the logical key, the physical address of the record is extracted from the index, and the record is read directly. Often, the index is maintained in logical key order. In that case, by following the index, key by key, it is possible to process such files sequentially. Searching an indexed sequential file is a two-step process: First search the index file using binary search or some other quick search technique; then search the record by direct access. Thus, the data retrieval from an indexed sequential file is quicker than that from a sequential file and slower than that from a direct file.

Finally, we discuss how data can be accessed from a randomly organized file. Various methods are available; one that is widely used is known as the division-remainder algorithm under the hashing method. Hashing refers to performing some calculations with the key and using the result of these calculations as the address of the record. The key must be numeric to perform the calculations. We now describe the algorithm below.

Let us suppose that n tracks are available in a track-based direct access storage device to store a randomly organized file that has a numeric key field that is to be used to determine the storage location of each record. Select the largest prime number p that is less than n. Let K be the value of the key field for a given record. Divide K by p and call the remainder r. Then r can have *only one* of p distinct values:

$$0, 1, 2, \ldots p - 1$$

Since $p < n$, each of the above numbers is also less than n; i.e., $r < n$. Then the record is stored in track number r. Thus, the algorithm associates a given record with a unique track, which determines the relative address of the record. If multiple records are stored on the same track, a chain of pointers is set up to access a designated record.

Mathematically, if a number K leaves a remainder r when divided by p, then so does every number of the form $K + \lambda p$, where λ is a nonzero number. Consequently, all records for which the key field has values K, $K + p$, $K + 2p$, $K + 3p$, ... will be stored on the same track; i.e., track r. This is usually called the *collision* problem in the division-remainder algorithm, since this situation arises when multiple records end up on the same track. As long as there is enough space on the track, these multiple records are stored on the same track and are connected by a chain of pointers. When the track becomes full, the records are stored on the overflow area and are accessed via pointers. Since the algorithm uses only p tracks out of the total of n tracks, the remaining $n - p$ tracks are available for the overflow area. When the overflow area becomes full, the whole file is reorganized and additional tracks are needed.

9.5 SELECTION CRITERIA FOR FILE ORGANIZATION METHOD

After all the files are designed and their respective access methods decided, the system team faces the task of selecting the organization method of each file. Although the file access method presupposes some specific organizations (e.g., a random access file cannot have a sequential organization), the selection may still be open. Davis ([1], pp. 388–390) has discussed a number of factors that may be considered as possible selection criteria:

a. **Activity.** Activity is a measure of the percentage of records that are actually accessed each time the file is processed. Files with high activity should be organized sequentially, since during each processing most of the records are accessed any way. Files with less activity should have random or indexed sequential organization.

b. **Response Time.** If response time is critical in an application involving on-line systems, it is probable that records will be accessed in a random, unpredictable order. Hence sequential files are inappropriate for such applications.

c. **Flexibility.** Flexibility refers to the option of accessing the records in a file sequentially or directly depending on the situation. Indexed sequential organization offers this choice. Also, multiple indexes can be created to allow for additional flexibility of file access.

d. **Volatility.** Volatility is a measure of the rate of additions to and deletions from a file. Files with high volatility are updated often, which makes the indexed sequential organization undesirable since with each update both the index file and the data file need to be updated. If highly volatile files have low activity, random organization should be selected. Otherwise, sequential organization is preferable.

e. **Total cost.** Total cost covers *all* types of direct cost such as physical storage cost, programming cost, execution cost, and indirect cost of user dissatisfaction. If poorly organized files cause unnecessary delays in producing reports or lead to high response times for on-line systems, there is a high indirect cost in the form of increased user dissatisfaction. In general, the objective is to select a file organization that allows the system to function at the lowest possible total cost. Usually, sequential files are cheaper to process and easier to handle through programs than random or indexed sequential files.

Let us close this section with the following comments of Davis ([1], p. 390):

These criteria are merely guidelines, not absolutes. They are the result of years of observation, trial, and error, and are in fact closer to folk wisdom than to science. Beginners often go overboard—the activity is high, and therefore the file must be organized sequentially. Don't forget that selecting a file organization is aimed at reducing the sum of several different costs, including the cost of storage, execution, programming, and usage. Selecting a file organization is perhaps the most important decision the analyst can make during detailed design. Take your time, and do it right.

9.6 SCHEMA DESIGN FOR A DATABASE

In a file-oriented environment, it may be necessary to design several direct access files with random or indexed sequential organization in order to support on-line systems. These files often have duplicate data elements, a situation that causes the problem of multiple file updates and data redundancy. In such cases the system team should seriously consider recommending a database environment. This presupposes the existence or procurement of a database management system (DBMS).

A DBMS is complex software that handles all access to a database. It allows the user to ask queries related to the data and to generate reports and/or graphics using those data. Conceptually, a database can be regarded as a collection of files with interrelated data elements. The DBMS automatically accesses these multiple files and retrieves the necessary data to meet the user's request. As a result, it saves a lot of programming effort on the part of the system team.

The details of database theory are given in Chapter 14. If you are unfamiliar with databases, we suggest you review that chapter before proceeding with the rest of this section. In this section we describe what is involved in a schema design process.

The conceptual format of data in a database constitutes the *schema*. It is often called the conceptual level of a database, whereas the actual data stored in a database are regarded as the internal level. A schema is a chart of the types of data that are used. It contains the names of the data elements, their formats, and their interrelationship. Hence the schema design is analogous to record and file designs.

The steps involved in schema design are as follows:

a. Design data elements at the record level (see Section 9.2).

b. Group these data elements into separate sets, each set corresponding to a file. Conceptually, these would be the files if you designed a file-based system.

c. Identify all data elements that are duplicated among the files.

d. Decide upon the data model (i.e., hierarchy, network, or relational) that you want to select for the system. This task in itself is a comprehensive job.

e. If you select a network or hierarchical model, then no duplicate data elements can exist among the records in the schema. So, design each record as the collection of nonduplicate data elements [see items (b) and (c)]. Then design the relationships among the records (called *sets* in CODASYL terminology) by using the duplicate data elements identified in (c).

f. If you select a relational model, design each relation consisting of all data elements belonging to the same set [see (b)]. Next, reduce the number of duplicate data elements to a minimum in such a way that two different relations should have at least one common data element. This duplication is needed since a relational database does not contain any entities corresponding to the "sets" in a network or hierarchical model.

This schema consists of the records and the sets in a network or hierarchical model and the relations in a relational model.

9.7 SPACE ESTIMATE FOR DATA FILES

Data files are stored on magnetic media, such as tapes and disks, which depend on the file access and file organization selected by the system team.

It is necessary to estimate how much auxiliary storage is needed in order to store all the data files as well as allow for future growth. Since storage on tapes is cheaper than on disks and since tapes are used primarily for backup, we shall consider space estimate procedures for disks only.

There are two principal mechanisms used in segmenting storage space on disks:

a. Sector-based or fixed-block architecture

b. Track addressed

Sector-based systems divide the surface of the disk into a series of fixed-length *sectors*. The *fixed-block architecture*, which is logically about the same as the sector-based storage, is used by many large computer systems. In a *track-addressed* system the disk space can be used more efficiently by adjusting the codes. Thus, it offers more flexibility for the programmer in disk usage. However, it requires careful planning during the record and file designs. The sector-based and the fixed-block architecture are somewhat easier to use than the track-addressed storage.

We now discuss the difference between a logical record and a physical record since the space estimate procedure concerns the logical as well as the physical records.

A *physical record* is the unit of data transferred between memory and auxiliary storage. A *logical record* is the unit of data used in the record format and is processed by a single iteration of a program. Code is written to process logical records, whereas hardware moves physical records. Although it is possible to store a single logical record as a single physical record, it is not necessary. The main objective of the space estimation procedure is to design the physical records so as to use auxiliary storage most efficiently.

The following is a step-by-step process to estimate the storage space for a single data file:

1. Compute the total number of characters per record.

2. Estimate the total number of records in the file, allowing for future growth.

3. Multiply the result in (1) by that in (2) to determine the total number of records in the file.

Next, we give two separate methods for space estimates, one for sector-based storage and the other for track-addressed storage.

9.7.1 Sector-Based Storage

On a sectored disk, a sector is the physical unit transferred between the disk and memory. If a logical record is smaller than a sector, then multiple logical records may be stored on a single sector. This phenomenon is known as *blocking*. With blocking, reading a single physical record provides multiple logical records. On the other hand, if a single logical record is larger than a sector, it is *spanned* over two or more sectors. With spanned records, accessing a single logical record involves reading two or more physical records.

The following procedure assumes that a logical record is smaller than a sector:

Let

$$S = \text{number of characters in a sector}$$

$$L = \text{number of characters in a record}$$

$$T = \text{number of records in the file}$$

Now use the following method:

1. Compute $M = \text{INT}(S/L)$, which represents the greatest integer not greater than the fraction S/L. Then M is the number of whole records that can fit on a single sector.
2. Compute T/M and round it upward to the nearest integer N. Then N is the number of sectors needed to store the file.

9.7.2 Track-Addressed Storage

On a track-addressed disk, logical records can be blocked together to form longer physical records. The number of logical records included in a single physical record is called the *blocking factor*. Data are stored on tracks in one of two possible formats: count/data format or count/key/data format. In either case, there exists a gap with no data between two consecutive physical records. Figures 9-3 and 9-4 illustrate these two formats. We note that a certain amount of unused space consisting of the track overhead, the count or count/key, and the gap is wasted in the sense that no data are recorded

TRACK OVER– HEAD	G A P	C O U N T	G A P	DATA	G A P	C O U N T	G A P	DATA	G A P	C O U N T	G A P
											..

FIGURE 9-3 Count/data format.

TRACK OVER-HEAD	G A P	C O U N T	G A P	K E Y	G A P	DATA	G A P	C O U N T	G A P	K E Y	G A P	DATA	..

FIGURE 9-4 Count/key/data format.

there. This "wasted space" must be taken into consideration in estimating the storage space for the data files. Each disk storage device manual indicates how many characters belong to this wasted space, which we call the *overhead* in our discussion.

The method of space estimate for track-addressed storage follows. We consider two cases according to whether the blocking factor is equal to 1 or greater than 1.

Case 1: Blocking Factor $= 1$

Let

H = overhead (i.e., track overhead, count or count/key, and gaps)

S = number of characters in a track

L = number of characters in a record

T = number of records in a file

1. Compute $P = H + L$ so P is the number of characters in each stored record that includes the overhead and the data.
2. Compute $M = \text{INT}(S/P)$; i.e., M is the greatest integer not greater than the fraction S/P. Then M is the total number of whole records that can be stored on a single track.
3. Compute T/M and round it upward to the nearest integer N. Then N is the total number of tracks needed to store the file.

Case 2: Blocking Factor > 1

Let B = blocking factor; i.e., $B > 1$, and let H, S, L, and T have the same interpretations as in case 1.

1. Compute $P = H + BL$ so P is the number of characters in each stored block that includes the overhead and the data.
2. Compute $M = \text{INT}(S/P)$ so M is the number of *whole* blocks that can be stored in a track.
3. Compute T/B and round it upward to the nearest integer U, so U is the number of whole blocks in the file.
4. Compute U/M and round it upward to the nearest integer N. Then N is the total number of tracks needed to store the file.

9.7.3 Size of Blocking Factor

We can reduce the wasted space on a track by blocking records together because by taking the blocking factor greater than 1 we eliminate the gaps between two consecutive logical records. In general, increasing the blocking factor leads to a reduction in the wasted space, but only up to a limit. Davis ([1], pp. 394–395) has shown by an example using the IBM 3330 disk pack that initially the impact of a larger blocking factor is dramatic in that by increasing the blocking factor from 1 to 5 the number of tracks used in the storage reduces from 80 to nearly 35. But then the situation levels off and eventually reaches a stage where a larger blocking factor results in a higher number of tracks required for storage. The exact relation between the blocking factor and the number of tracks needed for storage depends on the specific disk device and the record size. However, Davis suggests the following rule of thumb for deciding on the size of the blocking factor:

Draw a graph showing the number of tracks for storage as a function of the blocking factor. Select a blocking factor that lies near the beginning of the horizontal portion of the graph; i.e., when the graph starts to level off.

A second advantage of choosing the blocking factor to be greater than 1 is the increased *data transfer rate* between the auxiliary storage and the main memory. Thus, a magnetic tape drive reading data from a tape containing blocked records spends more of its time in the reading operation and less of its time in skipping over the empty tape space between two consecutive blocks. This results in a higher effective data transfer rate. In case of a magnetic disk, it is much faster to retrieve 10 records, say, stored sequentially as a block during one access than to retrieve them one at a time by means of 10 accesses.

To summarize, when the blocking factor is selected as greater than 1, two advantages result:

a. Less wasted space in auxiliary storage
b. Higher data transfer rate and lower access time

When records are blocked, an entire block must be transferred between the auxiliary storage and the main memory during a read/write operation.

Accordingly, the available memory size is a consideration in selecting the blocking factor.

9.7.4 An Example of Storage Estimate

To illustrate the procedures of Sections 9.7.1 and 9.7.2, let us suppose that a file contains 5000 records and each record has 125 characters.

(a) Sector-Based Storage

Here $S = 256$ (we assume this sector size)

$L = 125$

$T = 5000$

Therefore, $M = \text{INT}(S/L) = \text{INT}(256/125) = 2$

Hence two whole records can fit on a sector.

Next,
$$T/M = 5000/2 = 2500$$

Therefore, we need 2500 sectors to store the file.

(b) Track-Addressed Storage

Case 1: Blocking Factor = 1

Here $H = 200$ (we assume this overhead size)

$S = 18,660$ (we assume this track size)

$L = 125$

$T = 5000$

Now, $P = H + L = 200 + 125 = 325$

$M = \text{INT}(S/P) = \text{INT}(18660/325) = \text{INT}(57.415) = 57$

Hence 57 whole records can be stored on each track.

Finally,
$$T/M = 5000/57 = 87.719$$

Therefore,

$N = 88$, rounding T/M to the next higher integer

Therefore, we need 88 tracks to store the file.

Case 2: Blocking Factor = 16

Here
$$B = 16 \text{ (by assumption)}$$
$$H = 200$$
$$S = 18,660$$
$$L = 125$$
$$T = 5000$$

Therefore,

$$P = H + BL = 200 + 16 \times 125 = 2200$$
$$M = \text{INT} (S/P) = \text{INT}(18660/2200) = \text{INT}(8.482) = 8$$

Now,

$$T/B = 5000/16 = 312.5$$

Therefore,

$$U = 313$$

Finally,

$$U/M = 313/8 = 39.125$$

Therefore,

$$N = 40$$

Therefore, we need 40 tracks to store the file.
As expected, blocking reduces the storage space needed.

9.7.5 Automated Computation of Storage Space

Many computer vendors provide software that computes the total space needed for storing a data file. The user must provide all the necessary information for estimating the storage requirement. For example, in 1982 Data General introduced file management software called AOS INFOS II. It supported a utility called INDEXCALC that prompts users for the following items of information in order to compute the storage space for any AOS INFOS II file:

a. Number of keys in a record
b. Maximum length of the keys
c. Number of records per key
d. Maximum data record length

The output gives the number of characters that are estimated as the file size assuming a worst-case situation. Tracks or sectors are not used since Data General's AOS operating system allows records in a file to span disk sectors. By varying the maximum number of keys, it is possible to improve upon the space estimate since the latter assumes a worst-case scenario. Also, a program can be written to estimate the storage space for a file if there are too many files in the system.

9.8 AUXILIARY STORAGE ESTIMATE

The auxiliary storage estimate consists of the total space (number of tracks, sectors, etc.) needed for storing the following:

a. Data files
b. Program files
c. Input and output documents (if requested by the customer)

The space estimate for data files has been described in Sections 9.7.1 through 9.7.5. The system team estimates the space needed for each file then adds the amounts of space together to compute the total space needed to store all the data files used in the system.

The space estimate for program files is more difficult. If we are developing software for the first time, it is impossible to estimate the storage space for the software with a degree of accuracy similar to that in estimating storage for data files. But there are two exceptions to this general rule:

1. If the software to be developed is similar to software already implemented, the storage for the latter is a good estimate for the storage for the former. For example, if we already implemented a Data Entry/Edit subsystem for a financial application (say, Invoice, A/R) and if we know its storage requirement, that can be used as an estimate for a similar subsystem for a different application (e.g., Inventory Control). However, the programming language should be noted. (COBOL takes more storage than FORTRAN or PASCAL.)

2. If we are buying off-the-shelf software, the vendor can provide the storage requirement for the software. The vendor can also provide the memory requirement to run the package. Usually, the vendor provides two estimates for the memory:

☐ A minimum core memory to run the package
☐ An additional amount of memory to run it more efficiently; i.e., without too much page swapping, which will degrade performance and result in thrashing

Finally, let us consider the space estimate for input and output documents. Sometimes, due to legal requirements, the customer may want to store some or all of their input and output documents in electronic media. In that case, it is necessary to estimate the number of characters in the documents. The following quantification methodology is taken from Mittra ([2], pp. 26–27).

Data used in a system fall into two categories: input and output. Normally, the vehicle of input data is forms and that of output data is reports. The size of a form or a report is expressed as the number of bytes or characters in it. During the design phase not all of the forms and reports to be used in the system are known. Accordingly, the analyst takes the following sequence of steps to estimate the size of an average form or an average report:

a. Select a random sample of the available forms and reports.
b. For each form in the sample, count the number of characters per line and the total number of lines.
c. Use a simple average or a weighted average taken over all the forms in the sample. This determines the number of characters in an average form.
d. Repeat steps (a) through (c) for the reports in the sample to estimate the number of characters in an average report.

The estimates derived in (c) and (d) are, respectively, the *unit data object size* for input and output data.

The total volume of input or output data depends on the following factors:

1. **Frequency.** How often a form is entered or a report is generated.
2. **Unit size.** The unit data object size calculated in steps (c) and (d).
3. **Type.** Static or dynamic. If the data stays where they originate, their type is static; if the data originate in one location but are transmitted along a data communication line to another location, the data are dynamic. The dynamic type of data is used to calculate the traffic volume along a data communication line.
4. **Location.** The place of origination of the data. For static data, there is only one location; for dynamic data there are two locations, namely, the origin and the destination.
5. **Growth factor.** After a system is designed, the volume of both input and the output data will increase. The analyst should multiply the estimated data volume by the appropriate growth factor to determine projected needs.

Assume that the input data of a system consists of m forms, and that form i contains c_i characters and is input with a frequency f_i per month. Then the total projected *monthly input volume* V_{in} for the system is

$$V_{in} = \sum_{i=1}^{m} c_i f_i$$

Similarly, if there are n reports in the system, with report j containing d_j characters and being produced h_j times each month, the total projected *monthly output volume* V_{out} is given by

$$V_{out} = \sum_{j=1}^{n} d_j h_j$$

9.9 DATA COMMUNICATION NETWORK

If the system being designed is meant to be accessed by users located in different geographical areas, then proper data communication tools are needed. In an inventory control system, for example, the reorder level of an item is determined by comparing the reorder point setup in the optimal reorder policy with the actual inventory level obtained from the inventory master file. If the company has multiple warehouses and multiple sales facilities, the actual inventory level of an item must be determined by taking into account the current quantity on hand, adding to it the quantities received at the warehouses and subtracting from it the quantities sold. Since the warehouses and the sales facilities are located in different areas, a proper data communication network must be designed connecting all these places with the centralized data processing facility where the inventory master file is located. This involves a minimum of the following:

☐ Network topology showing the nodes where data are generated and arcs along which data are transmitted
☐ Network type (star, ring, tree structure, etc.)
☐ Speed of transmission (low, medium, or high)
☐ Mode of transmission (asynchronous, synchronous)
☐ Type of transmission (simplex, half-duplex, full-duplex)
☐ Line configuration (point to point, multidrop)
☐ Equipment (terminal, modem, channel, controller, multiplexer, concentrator, etc.)

Due to increasing uses of transaction-oriented processing supported by CRT terminals and microprocessors, the data communication network design is a major aspect of almost any system design. Even in batch-oriented systems, data (source documents, tapes, diskettes, etc.) are rarely sent via messenger or mail service. Instead they are transmitted electronically on a regular prescheduled basis (e.g., at 6 P.M. every day).

We now discuss briefly the items in the above checklist.

9.9.1 Speed of Transmission

The speed of transmission is measured by the number of bits per second (bps) sent over the line of the transmission. Three speeds are available:

a. Low speed—40–150 bps (e.g., telegraph lines)
b. Medium speed or voice band—300–14,400 bps (e.g., telephone lines); 9600 bps is average
c. High speed—over 9600 bps (normally used for computer-to-computer communication)

9.9.2 Mode of Transmission

Two modes are available: asynchronous and synchronous.

a. Asynchronous. One character at a time is transmitted with one START and one STOP bit, giving 10 bits/character in an 8-bit ASCII code.

b. Synchronous. Characters are transmitted as a group without START/STOP bits. The beginning and end of a character are determined by a timing mechanism within a modem.

9.9.3 Type of Transmission

Three types of transmission are possible: simplex, half-duplex, and full-duplex.

a. Simplex. Data transmission takes place in one direction only; e.g., a terminal can send data but cannot receive anything.

b. Half-duplex. Transmission is possible in both directions but can occur only in one direction at a time; e.g., POS terminals in transaction-oriented processing use half-duplex transmission. When such lines are used, the flow of transmission is stopped before the direction of transmission is reversed. The time to accomplish this is called *turnaround time,* which is usually 50–250 milliseconds.

c. Full-duplex. Transmission can occur in both directions simultaneously. No turnaround time is involved here. It is used for high-speed transmission lines, such as for communication between two computers.

9.9.4 Line Configurations

Two major configurations are used: point to point and multidrop or multipoint.

a. Point to point. Point-to-point configuration is a direct line between a terminal and the computer system. The terminal sends and receives data along a line that is not shared with any other terminal. It is used when the communication is almost continuous and requires fast response time. This is an expensive configuration since it needs two modems and a port in the controller for each terminal. See Figure 9-5.

The line can be *leased* or *switched*. A *leased line* is a permanent circuit connecting the terminal with the computer. A *switched line* is established through dial-up connection of the telephone line, and the terminal is then connected to the computer system.

b. Multidrop. Multidrop has more than one terminal on a single line connected to the computer system. When the line is used, only one

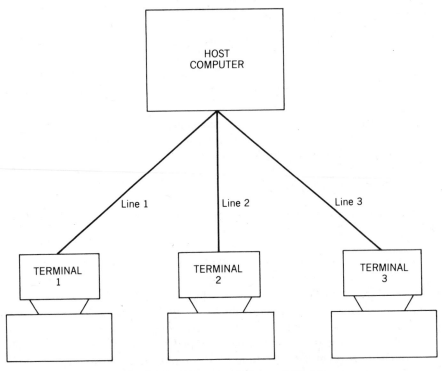

FIGURE 9-5   Point-to-point line configuration.

terminal can transmit to the computer system, but more than one terminal can receive data from the computer. This is less expensive than point-to-point lines. See Figure 9-6.

A leased line is mostly used for the multidrop configuration. Contact is established between the host computer and the terminal either by polling or by addressing. In *polling,* the system asks each terminal on the line if it has any data to send. The sequence for asking the terminals is determined on a prescheduled basis. If a terminal takes too much time in sending the data, it can face a programmed "timeout." In *addressing,* the system addresses the specific terminal that has a message to receive. See Figure 9-7 for a schematic overview of polling and addressing.

9.9.5 Network Types

There are three principal types of network:

a. *Star network* consists of one host computer connected to multiple terminals. A *pure* star network uses point-to-point lines between the host

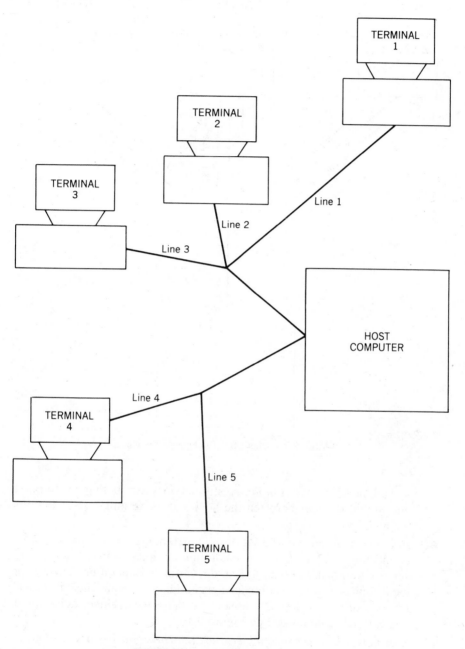

FIGURE 9-6 Multidrop line configuration.

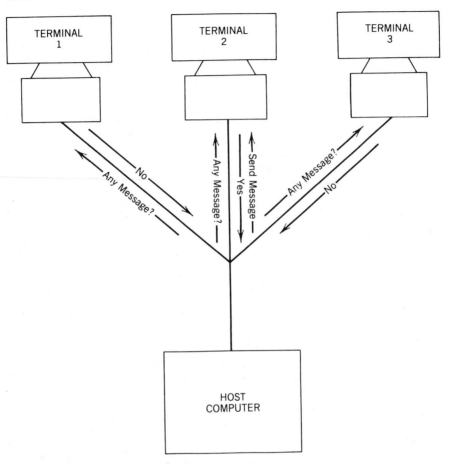

FIGURE 9-7   Schematic of polling/addressing.

and all the terminals. A *modified* star network uses both point-to-point and multidrop lines between the host and the terminals. See Figures 9-8 and 9-9.

b. *Ring network* consists of a series of minicomputers or microcomputers connected with one another, each mini or micro being connected to several terminals. It thus uses a decentralized operation, mostly for local communications within a single building or plant. Thus, it is ideal for local area networks. No host computer is used, since each mini or micro is self-contained. See Figure 9-10.

c. *Tree network*, also called a *hierarchy network*, consists of a host computer connected to two or more layers of subordinate computers. For example, the host computer may be located at the head office of a company and be connected to one layer of computers at the regional

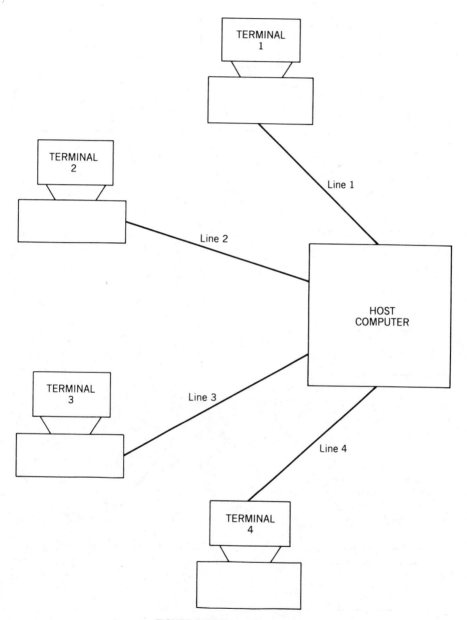

FIGURE 9-8 Pure star network.

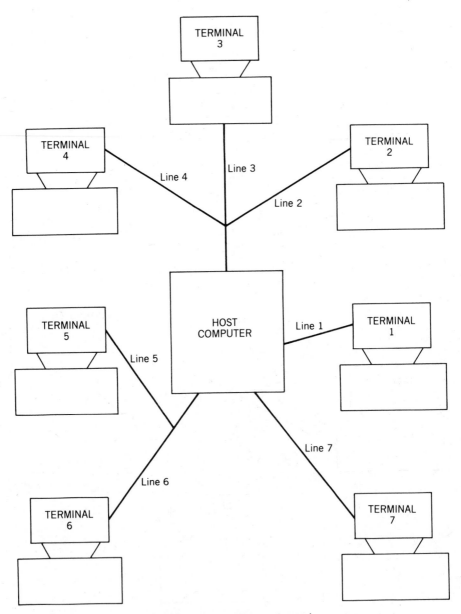

FIGURE 9-9 Modified star network.

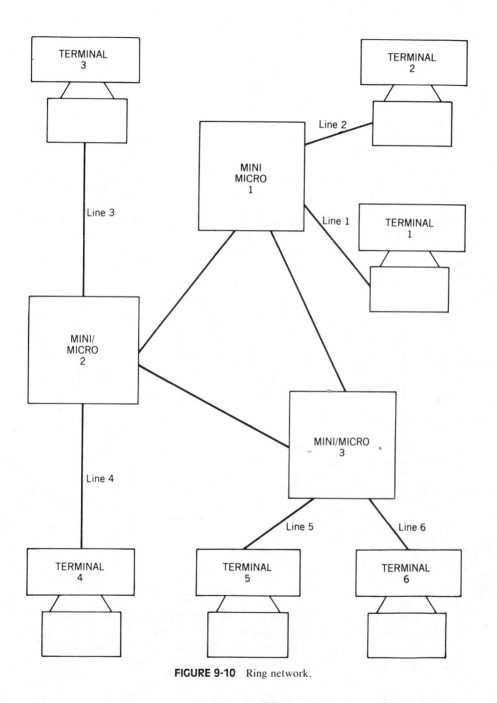

FIGURE 9-10 Ring network.

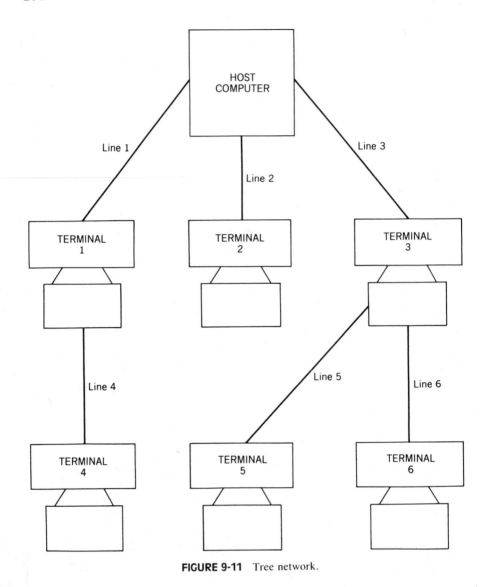

FIGURE 9-11   Tree network.

offices and a second layer of computers at the district offices of the company. Here we assume that several district offices report to a regional office and that all the regional offices report to the head office. See Figure 9-11.

9.9.6   Estimation of Traffic Volume

Any data communication network consists of multiple locations that must be connected with one another via data communication lines. Messages in the form of bits travel along these lines. It is necessary to estimate the volume of traffic along these lines in the form of number of bits transmitted per second. If there are n locations in the network, the traffic volumes can be represented by an $n \times n$ matrix:

$$T = (t_{ij}) \qquad i, j = 1, 2, 3, \ldots, n$$

where t_{ij} = number of bits transmitted per second between the locations i and j

We can use the computational procedure described in Section 9.8 to estimate the value of each t_{ij}. Here the data type is *dynamic* since the data moves from the origin to the destination. Each t_{ij} is computed by taking into consideration the frequency of transmission and the unit data object size.

9.9.7 Network Topology

A *network topology* consists of a set of nodes connected by a set of arcs. The nodes represent computers, terminals, or some type of communication control units in various locations, and the arcs are the communication lines along which data are transmitted from one node to another. The lines can be dedicated or switched. The topology structure is represented as a connected graph. In addition, detailed specifications must be provided about the type, speed, and mode of transmission for each line, as well as the line configurations.

Figure 9-12 shows a network topology. It contains two separate network types: a star network, consisting of a host computer connected to multiple terminals via multidrop lines, and a ring network, consisting of four minicomputers that are connected to terminals or other I/O devices.

The design of a network topology is a complicated job. The goal is to achieve a specified performance at a minimal cost. The problem can be characterized as follows (see [4], pp. 33–34):

Given
    Locations of the hosts and terminals
    Traffic matrix (t_{ij})
    Cost matrix (c_{ij})
Performance constraints
    Reliability
    Delay/Throughput
Variables
    Topology
    Line capacities
    Flow assignment
Goal
    Minimize cost

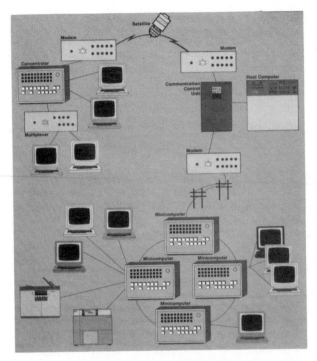

FIGURE 9-12   Example of network topology. (Reproduced with permission from G.B. Shelley and T.J. Cashman, *Introduction to Computers and Data Processing*, Anaheim Publishing Company, Brea, CA, 1980.)

The hosts and terminals are the ultimate producers and consumers of information. For most design purposes, hosts and terminals are equivalent, and it will often be convenient to lump them together under the term *location* or *site*.

The traffic matrix (t_{ij}) estimates the number of bits to be transmitted per second (see Section 9.9.6). If t_{ij}'s cannot be estimated due to lack of information, the following heuristic is used:

$$t_{ij} = k \times \frac{(\text{Data volume at node } i) \times (\text{Data volume at node } j)}{\text{Distance between nodes } i \text{ and } j}$$

The cost matrix (C_{ij}) tells us how much the communication lines of various speeds between the nodes i and j will cost. In general, the cost C_{ij} depends on the distance between i and j and on the speed of the line connecting them in a highly nonlinear fashion.

Just to complicate things further, only a discrete set of speeds is available; for example, 50, 110, 300, 600, 1200, 2400, 4800, 9600, 56,000 and 230,400 bps. As a result, if the optimum speed for a line turns out to be, say, 20,000 bps, the line will not be available. Either a 9600-bps line or a 56,000-bps line will have to be used, the former providing too little performance and the latter too high a cost. The effect of "rounding off" all the speeds to the set of available values may produce a configuration that is far from optimal. In order to solve the problem optimally in a comprehensive manner, we have to use the methods of integer linear programming, which is beyond the scope of this book.

9.9.8 Data Communication Software

The data communication functions are handled by the communication software that works in conjunction with the communication hardware consisting of the host computer, terminals, communication lines, modems, concentrators, multiplexers, and so on. The software does the following:

☐ Processes data received over the network
☐ Sends messages back to terminals
☐ Controls the network
☐ Establishes contact with terminals
☐ Processes any line errors
☐ Performs polling and addressing of terminals for multidrop lines

Various equipment manufacturers provide different software packages for these purposes.

We close this section with the following comments by Lucas ([3], p. 204):

All the options must be considered by the network designer. It is important for the systems analyst to work with the communications specialist to determine where various functions are to be performed. We may want to have some data editing done at the terminal or store screen formats there to reduce the load on the host CPU and improve response times. The options available for communications networks contribute a great deal of flexibility to the systems design task.

9.10 EQUIPMENT SPECIFICATIONS

The equipment specifications relate directly to the selection of hardware that is necessary for implementing the system. The types and quantities of the hardware needed are specified in generic terms. We can include the following items as essential hardware:

Processor

CPU
Main storage

Storage

Tape
Disk
Mass storage device
Diskette

Peripherals

Serial and/or line printer
Card reader punch
Tape drive
Disk drive
Terminal

Communication Hardware

Channel
Modem
Concentrator
Statistical multiplexer
Controller

The term *grounding* is sometimes used to indicate the process whereby a specific piece of equipment is prescribed in order to address a conceptual system component as determined from the analysis of a given user need. The basis of grounding is the storage estimate data. The detailed specification of hardware and communication equipment results from a series of grounding processes.

In order to prepare the equipment specifications, the system team must review the state-of-the-art hardware and software technology. Sources for hardware include the following:

   a. Service bureaus
   b. Timesharing
   c. Remote batch processing
   d. Inhouse computer equipment

Similarly, sources of software include:

   a. Software houses
   b. Off-the-shelf software packages
   c. Inhouse expertise

	Advantages	Disadvantages
Hardware sources		
Service bureau	Professional service	Costly, lack of control
Timesharing	No maintenance	Variable cost, external management
In-house computer	Complete control	Maintenance responsibility
Software sources		
Software house	Variety of expertise	Costly, external management
Off-the-shelf software package	Variable	Customization needed
In-house expertise	Complete control No extra expenses	May be inadequate for the job

FIGURE 9-13 Comparison of advantages and disadvantages of hardware and software sources.

Possible advantages and disadvantages of the above sources are summarized in Figure 9-13.

The system team must establish selection criteria such as capabilities, ease of use, vendor support, training, and cost (purchase, rental, or lease) to decide upon specific equipment. In order to get preliminary information, the team prepares one or more requests for proposal (RFPs) and invites vendors to bid on them. Normally, an RFP lists the requirements on each piece of equipment as either mandatory or desirable. The former are minimum requirements that must be satisfied and the latter are enhancements on the minimal configuration.

After receiving quotations from the vendors, the team uses its selection criteria to narrow down the list of prospective vendors to whom requests for quotation (RFQs) will be sent. An RFQ is much more specific than an RFP and as such uses definitive requirements. Usually, RFPs are used at the initial stage when the procurement is open, and RFQs are sent at the final stage to a much smaller group of vendors. Figures 9-14 and 9-15 give examples of an RFP and an RFQ.

The final selection of vendors for the equipment depends upon the total cost quoted by the vendors. Often the system team distinguishes between two types of cost considerations, hard dollars and soft dollars. *Hard dollars* refer to the actual money needed to procure the equipment meeting the mandatory requirements. Thus, hard dollars are quantifiable. *Soft dollars*, on the other hand, represent subjective assessments of the economic worth of a desirable feature or capability of the equipment. Soft dollars represent a mechanism to select vendors who are not the lowest bidders in respect of hard dollars but who can provide features or capabilities above and beyond

TOY WORLD, INC.
REQUEST FOR PROPOSAL (RFP)

1. Introduction
   – Objectives of RFP
   – Basic Functional Capabilities
   – Background Information
2. Required Applications
   – Order Processing
   – Inventory Control
   – Customer Billing
3. System Specifications
   – Hardware Configuration
   – System Software
   – Languages and Packages
4. Contractual Support by Vendor
5. Evaluation Criteria
   – Robustness of System
   – User Training and Documentation
   – Modularity of System
   – Hard Dollar Value
   – Soft Dollar Value

FIGURE 9-14   Example of an RFP.

the mandatory requirements. Vendors should be notified in advance as to what soft dollar considerations will be given to their proposals.

The end product of the equipment specification process is a detailed listing of the various pieces of equipment needed by the system with their respective capabilities. To facilitate understanding, several diagrams should be prepared showing the network topology, locations of the communication equipment, and conceptual layouts of other hardware at respective locations.

Figures 9-16 and 9-17 represent a sample equipment configuration and a supporting data communication network.

TOY WORLD, INC.
REQUEST FOR QUOTATION (RFQ)

Four Microprocessors
Specifications:
1. IBM PC Compatible
2. 640 K RAM
3. 20 MB Hard Disk
4. Color Monitor
5. 1920-character (24 × 80) Display Screen
6. Serial Printer Interface
7. 10-key Function Keypad
8. Built-in Modem 1200 Baud

FIGURE 9-15   Example of an RFQ.

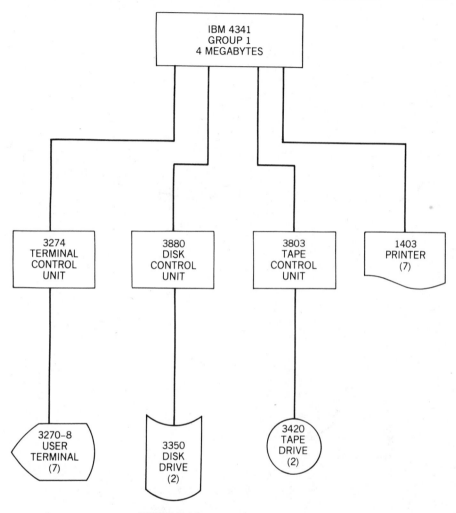

FIGURE 9-16 Equipment configuration.

9.11 PERSONNEL SELECTION FOR IMPLEMENTATION AND MAINTENANCE

By now the system team has a complete layout of the physical system in the form of input/output requirements, file or database structures, estimate of auxiliary storage, data communication configuration (if applicable), and equipment specifications. The next and the final phase is the system implementation, maintenance, and evaluation. We shall discuss this in Chapter 11. However, in order to make that phase possible, it is necessary to select the appropriate personnel. Thus, the final task during the detailed design

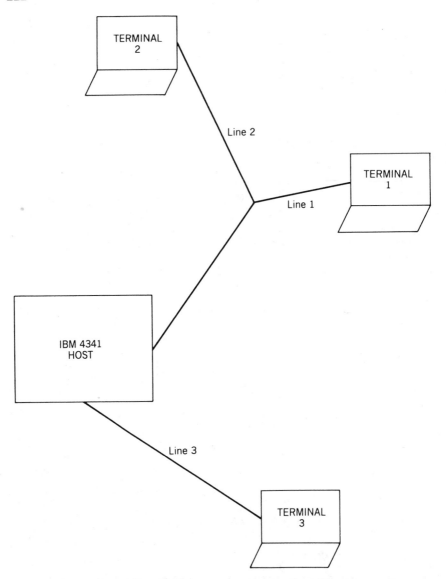

FIGURE 9-17   Data communication network.

phase is to identify the personnel requirements to implement and subsequently maintain the system. The emphasis in this selection process is on programming capability, system operation, and system troubleshooting in general. The system development process is now ready to convert the "paper" system into an "electronics" system. We need the right type of personnel to do this job.

The typical criteria for personnel selection include the following:

1. Education and/or training
2. Experience
3. Attitude
4. Past employment record
5. References

In most cases, job descriptions are also used. In addition, due to legal requirements, the affirmative action policy is enforced by most organizations. This policy is mandatory for all companies that have federal government contracts. It means that if one candidate is selected over others, that selection must be justifiable if audited by an affirmative action officer or the Equal Employment Opportunity Commission. Such justification can be particularly difficult if a job vacancy has been nationally advertised and several hundred applicants have responded.

The personnel selection process is essentially subjective in nature. Quantification of the selection criteria is rarely done and perhaps is not quite feasible. Wetherbe ([5], pp. 174–178) has described a quasi-quantitative selection technique that may be called a *score analysis process*. It consists of the following steps:

a. Identify attributes that are meaningful for the position, and categorize them by the five criteria listed above. There may be multiple attributes for a category.
b. Rank the attributes in descending order of priority relevant to the job by assigning weights to each attribute.
c. After interviewing a candidate and examining his or her credentials, assign a raw score to each attribute, multiple that score by the weight assigned to the attribute, and add the values together to get the total score.
d. Repeat step (c) for each candidate. Finally, make a list of all the candidates with their respective scores.
e. Select the candidate with the highest score.

Although somewhat quantitative in nature, the method still uses subjective judgement in assigning weights to the attributes and raw scores for each applicant on these attributes.

9.12 AUTOMATED METHOD FOR ESTIMATING HARDWARE, SOFTWARE, AND PERSONNEL

In 1982, Howard A. Rubin, a professor of computer science at Hunter College, developed a software called *Estimacs* for estimating the time, cost, staffing, hardware, and cumulative resource needs and risks for mainframe system development projects.

The program is based on answers to 25 questions about the user organization and the general size, sophistication, and complexity of the target system, expressed in business terms. It allows alternative estimates to be developed through refinements and analysis of design tradeoffs through a series of what if questions.

Estimacs currently has four modules:

a. *Software development estimator,* which estimates software development efforts in terms of total hours, using several input parameters including information on the developer's knowledge of the project, complexity of the customer organization, and general size and sophistication of the final system. Output information covers total work effort in hours, phase distribution of the work over the software life cycle, productivity tracking, and the relationship of each answer to the estimate.

b. *Staffing and cost estimator,* which determines the personnel needs at each skill level over the software project life cycle.

c. *Hardware estimator,* which estimates the hardware configuration requirements in terms of CPU power and I/O components.

d. *Portfolio estimator,* which produces the cumulative resource needs, related cost, and relative risks of undertaking too many concurrent projects.

In each of these packages, multiple iterations are possible by juggling various inputs to produce the most viable program for the user. The system is in use at Warner Lambert, a manufacturer; Peat, Marwick Mitchell, an accounting and consulting firm; and Manufacturers Hanover, a bank. Estimacs is available commercially through Management and Computer Services Inc. (Valley Forge, PA), a diversified EDP organization.

Estimacs uses color graphics to demonstrate relationships between inputs and forecasts on Apple II and IBM Personal Computers. All four Estimacs modules are available for $20,000; modules can also be purchased individually. Corporate multiple copy licenses are available.

9.13 END PRODUCT OF DETAILED SYSTEM DESIGN PHASE

The end product of the detailed system design phase is a report containing complete specifications of the system level design. Since the end product of the preliminary system design phase contains complete layouts of the input and output documents (see Section 7.11), it is not necessary to repeat them here. However, a brief mention should be made of them, and some sample input and output documents should be included. Accordingly, the following list provides a tentative table of contents of the detailed design report:

1. **Data Flow Diagrams.** Include the level 1 data flow diagram for the entire system and the most detailed micro level data flow diagram for each subsystem. This reinforces the progression from the logical system to the physical system and refreshes the customer's memory with the requirements analysis of the proposed system.

2. **Functional Capabilities.** Describe in brief narrative form the functional capabilities of the whole system and of each subsystem. Correlate these descriptions with the data flow diagrams and identify the interfaces among the subsystems.

3. **System Flowcharts.** Include one system flowchart for each data flow diagram given in section 1 of the report. Identify the transition from the logical to the physical system by preparing a cross-reference list showing

   a. Which program in a system flowchart corresponds to which process(es) in the corresponding data flow diagram that it replaces

   b. Which file in a system flowchart corresponds to which data store(s) in the corresponding data flow diagram

4. **Samples of Input and Output Documents.** Include some samples of input documents, output reports, and screen displays (if applicable). This is a carryover from the preliminary system design phase.

5. **File or Schema Structures.** For a file-oriented system, describe the format of each file by describing its record structure in the following form:

   a. Name of each field

   b. Description of each field

   c. Format of each field (numeric, alphanumeric, etc.), and length

   Also, describe the file access and organization methods.

   For a database-oriented system, describe the schema structure. Include a diagram of the schema if you are using a hierarchical or network model.

6. **Auxiliary Storage Estimate.** Give an estimate of the total auxiliary storage needed for the system. Include storage space for all data files, program files, and input/output documents (if required by the customer). Include the detailed computation of the space estimate for one data file as an example to describe the basis of your estimates.

7. **Equipment Specification.** Give a detailed list of all the hardware needed for the system. Identify which of the pieces of equipment has to be procured, since the procurement process is usually lengthy. Include an equipment configuration chart similar to Figure 9-16.

8. **Data Communication Network (if needed).** Describe the network configuration required by the system. Give an estimate of the traffic

volume along the data communication lines. Also, include the specifications listed in Section 9.9. Attach a network topology diagram similar to Figure 9-17.

9. **Personnel Estimate.** Give a list of personnel needed for implementation and maintenance of the system. Justify why they are needed.
Include a tentative implementation schedule for the system assuming
that the required personnel will be available. This schedule will be
followed during the next phase of system implementation.

9.14 SUMMARY

This chapter described how the detailed system design is done. As a sequel
to the preliminary system design phase discussed in Chapter 7, the detailed
system design explores the following seven topics:

a. Data elements and record structure
b. File organization
c. Schema structure
d. Auxiliary storage estimate
e. Data communication network
f. Equipment specification
g. Personnel selection for system implementation and maintenance

Using the data flow entries in the data dictionary, we design the complete
set of data elements for the system. Conceptually, a data flow can be regarded
as a collection of data elements. Hence we analyze each data flow entry and
expand it until we reach the data elements to be used in the physical system.
We then assign a *unique* name to each data element and decide on the format,
description, and location of each. The collection of all data elements pertaining to a specific topic forms a record, which may be of fixed length or
variable length. Finally, we group together all these records to constitute a
file. The contents of each file must be determined from the corresponding
data stores described in the data dictionary.

A file has three possible organizations: sequential, random, or indexed
sequential. The organization of a specific file is determined by considering
several factors such as its activity, response time, flexibility, volatility, and
processing cost. Sequential files can be accessed only according to the order
of the records in the file: a record in a random file can be accessed directly.
An index sequential file uses a two-step process: It first determines the location of the desired record by searching an index file using binary search;
then it accesses the record directly in the data file.

With the advent of numerous database management systems available on all types of computers from mainframes to micros, often a database is regarded as a viable alternative to a file-oriented system. If this option is chosen, we must first decide the data model (hierarchy, network, or relational) and then design the schema structure, which is somewhat analogous to the structure design of files. However, in a database the redundancy of data is either eliminated (network or hierarchy data model) or minimized (relational data model).

The auxiliary storage estimate consists of estimating the space needed in the form of number of sectors or number of tracks for storing the data files, the program files, and the input and output documents (if so requested by the customer). Two separate methods are given here, one for sector-based storage and the other for track-based storage. For the latter, two cases are possible, depending on whether the blocking factor of the file is 1 or greater than 1. Some vendors provide an automated computation method to estimate the storage space.

If the physical system is distributed over multiple locations, it is necessary to have a data communication network for electronic transmission of data. An appropriate network topology must be designed to transmit data subject to minimum or near-minimum cost. The network specifications must include, as a minimum, the network type, the line configuration, and the speed, mode, and type of transmission. Four common types of network are pure star, modified star, tree, and ring.

The equipment specification consists of a complete list of all hardware needed. It must include the central processing unit, main storage or memory, auxiliary storage, and peripheral devices such as tapes, disks, printers, and terminals. If there is a data communication component of the system, additional data communication hardware such as modems, communication channels, multiplexers, and concentrators should be included. Normally, we use the mechanisms of request for proposals and request for quotations to select the vendors for procuring the necessary equipment on the basis of competitive bidding. In order to decide on one or more vendors, we should consider both hard dollars and soft dollars to estimate the value of the proposed hardware.

The final step in the detailed system design is the selection of personnel needed for successful implementation and maintenance of the system. Here special emphasis is placed on programming and operational capabilities, since they are critical for converting the "paper" system into an "electronics" system. The personnel selection is essentially subjective in nature, although some companies use a quasi-quantitative score analysis process.

A software called Estimacs, introduced by Howard A. Rubin in 1982, provides an automated process of estimating the hardware, software, staffing cost, and relative risk of undertaking too many concurrent projects.

The chapter closed with an outline of the contents of the report that forms the end product of this phase.

9.15   KEY WORDS

The following key words are used in this chapter:

auxiliary storage
blocking factor
central processing unit (CPU)
concentrator
count/data format
count/key/data format
data communication
data dictionary
data element
data flow
data model: hierarchy
data model: network
data model: relational
data store
database
detailed system design
direct or random file
division/remainder algorithm
dynamic data type
Estimacs
file access
file activity
file flexibility
file organization
file structure
file volatility
fixed length record
grounding
growth factor
hard dollar
hashing method
index file

indexed sequential file
line configuration (point to point, multidrop)
logical record
mode of transmission (synchronous, asynchronous)
modem
multiplexer
network topology
network type (star, tree, ring)
physical record
record structure
request for proposal (RFP)
request for quotation (RFQ)
response time
schema
score analysis
sector-based storage
sequential file
soft dollar
space estimate
speed of transmission (low, medium, high)
static data type
terminal
track-addressed storage
traffic volume
type of transmission (simplex, half-duplex, full-duplex)
unit data object size
variable length record

REFERENCES

1. William S. Davis, *System Analysis and Design: A Structured Approach*, Addison-Wesley, Reading, MA, 1983.

2. Sitansu S. Mittra, "Quantification: An Essential Part of System Design," *Journal of Systems Management,* June 1983, pp. 26–28.

3. Henry C. Lucas, Jr., *The Analysis, Design, and Implementation of Information Systems,* McGraw Hill, New York, 1985.

4. Andrew S. Tanenbaum, *Computer Networks,* Prentice-Hall, Englewood Cliffs, NJ, 1981.

5. James C. Wetherbe, *Systems Analysis for Computer-Based Information Systems,* West Publishing Company, St. Paul, MN, 1979.

REVIEW QUESTIONS

1. How does the detailed system design differ from the preliminary system design?

2. List the topics that are included in the detailed system design phase.

3. Explain the meaning of the statement, A data flow usually represents data elements at a group level.

4. Distinguish between a fixed length record and a variable length record. Comment on the advantages and disadvantages of each.

5. What is meant by file access? How many types of file access are there? Name them.

6. How many types of file organization are there? Explain each one of them.

7. How are file access and file organization related to each other? How do you decide on a specific file organization method?

8. What is a blocking factor? Is it true that the larger the blocking factor the greater the reduction in wasted storage space? Why?

9. Suppose that a customer wants to store all input and output documents in magnetic media devices such as tapes. How would you estimate the size of an "average" document?

10. What items should be included in the specifications of a data communication network configuration? Comment briefly on each item.

11. How would you estimate the values of each cell t_{ij} of a traffic matrix?

12. Explain the distinction between hard dollars and soft dollars. Comment on their relationship with tangible versus intangible benefits.

13. What are RFP and RFQ? How are they used for equipment specification?

14. Why do you need to select personnel at the end of the detailed design phase? How does this selection process affect the personnel already working on the team?

15. Explain the process of score analysis to select personnel.

Detailed System Design: Case Studies

10.1 INTRODUCTION

In this chapter we shall discuss the detailed design phase of the two case studies—Toy World, Inc., and Massachusetts Educational Foundation. It thus provides a continuation of the work done in Chapter 8. For each case study, we start with system flowcharts developed in Chapter 8 and complete the system level design by following the theoretical framework developed in Chapter 9.

10.2 DETAILED SYSTEM DESIGN FOR TOY WORLD

10.2.1 Selection of a Subsystem for Detailed Design

Section 8.2.2 contains six system flowcharts (see Figures 8-2 through 8-7) for the complete system. These correspond to six processes that comprise the three subsystems:

☐ Order processing
☐ Inventory control
☐ Sales management

(see Section 4.5 for more details). The order processing subsystem is described by the three system flowcharts given in Figures 8-2, 8-5, and 8-6.

HPC now selects the system flowchart in Figure 8-5 and provides detailed specifications of the order processing subsystem built around the process of implementing customer requested changes. This gives a flavor of the detailed system design methodology and can be repeated for other subsystems in a similar way.

Figure 10-1, which is the same as Figure 8-5, gives the system flowchart of the selected process. This selected process is used if it becomes necessary to change a customer's transaction record. For example, the following fields within this record can be changed or altered:

1. Customer's transaction number, date, or code
2. Customer's name, address, or phone number
3. Cancellation of an order
4. Alteration of an order
5. Shipment date
6. Invoice number
7. Salesperson's number
8. District
9. Substitution of an order

HPC has chosen this particular process within the order processing subsystem because it demonstrates the subsystem's inquiry process, as well as its effect on four major files, namely master file, sales file, customer file, and inventory file.

The primary purpose of the process is to allow for varying inquiries, changes, or alterations within the subsystem itself. If, for example, a customer decides to increase an order, it is possible for a clerk to access the inventory file to ascertain the status of the requested style. If the style is in stock, the clerk can process the customer change transaction immediately. If, however, the requested style is not on hand, the request is placed on backorder or a substitution order is processed. In effect, all four files are updated.

10.2.2 File Designs

HPC now provides the detailed record format; that is, the name, description, and data format of each field within a record, the file access, and the file organization method for each of the four files: master file, sales file, customer file, and inventory file.

Figures 10-2 through 10-5 provide the detailed formats of the four files.

Figure 10-1 shows that all of these files have random access and are implemented as disk files. Since the processing of customer requested changes occurs on-line (see Figures 8-15 and 8-18 through 8-24), each file is organized

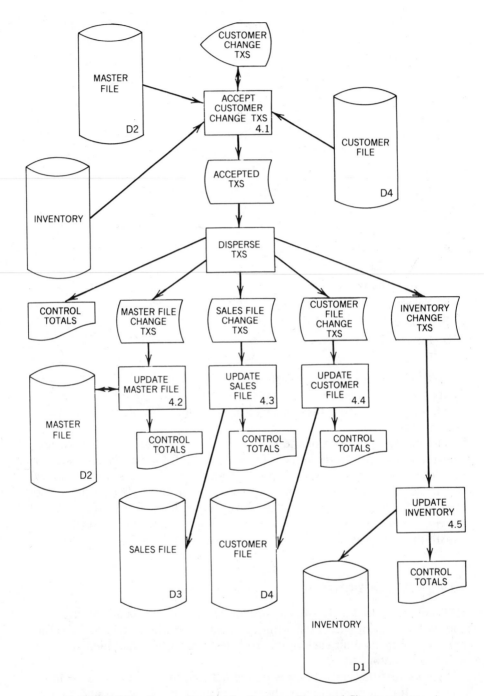

FIGURE 10-1 System flowchart for Customer Requested Change process.

```
FILE:  MASTERFILE
MASTERFILE-RECORD
```

NAME	DESCRIPTION	SIZE AND FORMAT
MF-invoice-order-no	Key field that uniquely describes a customer order	s9(8) numeric packed
MF-customer-no	Number that uniquely identifies a customer	s9(6) numeric packed
MF-customer-name	Identified customer	x(16) alphanumeric
MF-customer-address	Describes location of customer business	Group level
...MF-customer-street	Identifies street location of customer	x(15) alphanumeric
...MF-customer-city-state	Identifies city and state of customer	x(20) alphanumeric
...MF-customer-zip	Identifies zip-code in a customer's mailing address; defined by USPS	s9(5) numeric packed
MF-date-order-received	Date order was received by salesperson	s9(7) numeric packed
MF-date-shipment-requested	Date that customer would like shipment of order	s9(7) numeric packed
MF-salesperson-no	Uniquely identifies a salesperson	Group level
...Filler	—	x(4) alphanumeric
...MF-district	Uniquely identifies district a salesperson works in	x(1) alphanumeric
.....MF-district-0001	88 level	A
.....MF-district-0002	88 level	B
.....MF-district-0003	88 level	C
.....MF-district-0004	88 level	D
.....MF-district-0005	88 level	E
.....MF-district-0006	88 level	F
.....MF-district-0007	88 level	G
MF-total-cost	Total cost of customer order	9(5) ⋅ 99 packed

FIGURE 10-2 Format of MASTER FILE.

NAME	DESCRIPTION	SIZE AND FORMAT
MF-total-entries	Indicates total number of styles per order	s9(2) numeric
MF-order-entries	Contains fields that indicate a style entry occurrence	Group level—occurs 10 times depending on MF-total-entries
...MF-style-category	Identifies various style categories such as Christmas, Easter, Valentine	x(2) alphanumeric
...MF-style-no-ordered	Identifies style number ordered per entry occurrence	s9(5) numeric packed
...MF-bin-location	Identifies location of inventory per entry occurrence	x(7) alphanumeric
...MF-quantity-ordered	Identifies number of a particular style ordered per order entry occurrence	s9(5) numeric packed
...MF-qty-shipped	Identifies the number of a particular style shipped to customer per order entry occurrence	s9(5) numeric packed
...MF-total-price	Total price per order entry occurrence	9(5) . 99 packed
...MF-entry-status	Flag that indicates if occurrence has been filled or not	x(1) alphanumeric
	a = MF-not-processed	
	b = MF-order-filled	
....MF-not-processed	88 level	A
....MF-order-filler	88 level	B
...MF-order-status	Describes status of order per occurrence	x(1) alphanumeric
	A = MF-A-OK	
	B = MF-full-substitution	
	C = MF-partial-substitution	
	D = MF-full-back-order	
	E = MF-partial-back-order	
	F = MF-full-cancellation	
	G = MF-partial-cancellation	

`.....MF-a-ok`	88 level	A
`.....MF-full-substitution`	88 level	B
`.....MF-partial-substitution`	88 level	C
`.....MF-full-back-order`	88 level	D
`.....MF-partial-back-order`	88 level	E
`.....MF-full-cancellation`	88 level	F
`.....MF-partial cancellation`	88 level	G
`...MF-substituted-style-no`	Number that describes which style replaced the style ordered in this occurrence	s9(5) numeric packed
`..MF-qty-substituted`	Number of boxes of a style that was substituted in an occurrence	s9(5) numeric packed
`...MF-subs-bin-location`	Identifies location of the substituted order per occurrence	x(7) alphanumeric
`...MF-qty-back-ordered`	Identifies quantity of style on back-order per occurrence	s9(5) numeric packed

FIGURE 10-2 *continued*

FILE: SALES FILE

SALES-RECORD

NAME	DESCRIPTION	SIZE AND FORMAT
SA-district	Code that uniquely identifies the sales district	x(4) alphanumeric
SA-style-category	Identifies category of card style	x(2) alphanumeric
SA-style-no	Identifies card style by number	s9(5) numeric packed
SA-style-name	Identifies name of card style	x(15) alphanumeric
SA-total-gross	Indicates total gross sales per division	9(9) ⌄ 99 packed
SA-unit-price	Identifies price per box	9(2) ⌄ 99 packed
SA-qty-per-box	Identifies quantity of cards per box	s9(4) numeric packed
SA-qty-sold	Indicates quantity of boxes sold	s9(9) numeric packed
SA-qty-canceled	Indicates quantity of boxes canceled	s9(9) numeric packed
SA-qty-returned	Indicates quantity of boxes returned	s9(9) numeric packed

FIGURE 10-3 Format of SALES FILE.

as an indexed sequential file. The following table gives the index or indices for each file:

File name	Index Name(s)
Master file	MF-invoice-order-no
Sales file	SA-district, SA-style-no
Customer file	CU-number
Inventory file	IN-style-no

The programs that maintain the file update part of the order processing subsystem are

1. CRC 410, which accepts and edits any changes or alterations to the subsystem. This program allows the process to be continuously used.

FILE: CUSTOMER FILE

CUSTOMER-RECORD

NAME	DESCRIPTION	SIZE AND FORMAT
CU-number	Identifies a customer by number	s9(6) numeric packed
CU-name	Identifies a customer by name	x(16) alphanumeric
CU-address	Identifies a customer by address	Group level
...CU-street	Identifies a customer street	x(15) alphanumeric
...CU-city-state	Identifies a customer city and state	x(20) alphanumeric
... CU-zip	Identifies a customer zip code	s9(5) numeric packed
CU-telephone	Identifies a customer telephone number	Group level
...CU-area-code	Identifies the area code of customer telephone number	s9(3) numeric packed
...CU-exchange	Identifies the first three digits of customer telephone number	s9(3) numeric packed
...CU-phone	Identifies the last four digits of customer telephone number	s9(4) numeric packed
CU-balance	Identifies the customer balance amount	9(9) ⌄ 99 packed

FIGURE 10-4 Format of CUSTOMER FILE.

2. CRC 420 disperses the new information to one or more of the major files; it is used for batch processing.
3. CRC 430 updates the Master File.
4. CRC 440 updates the Sales File.
5. CRC 450 updates the Customer File.
6. CRC 460 updates the Inventory File.

Figures 10-6 through 10-10 show samples of Control Totals reports generated by the five programs CRC 420, CRC 430, CRC 440, CRC 450, and CRC 460, respectively. These reports are used to ensure quality control of the system.

FILE: INVENTORY FILE

INVENTORY-RECORD

NAME ·	DESCRIPTION	SIZE AND FORMAT
IN-style-category	Identifies various style categories and bin locations	X(2) alphanumeric
IN-selection-code	Identifies style selected	x(1) alphanumeric
IN-unit-price	Identifies price per box	9(2) · 99 packed
IN-style-no	Identifies card style by number	s9(5) numeric packed
IN-style-name	Identifies card style by name	x(15) alphanumeric
IN-bin-location	Identifies bin location of a card style	x(7) alphanumeric
IN-qty-on-hand	Number of boxes in a warehouse of this style	s9(9) numeric packed
IN-qty-per-box	Number of cards per box of this style	s9(4) numeric packed
IN-back-order	Number of boxes of this style ordered but not received at warehouse	s9(9) numeric packed
IN-back-order-due-date	Date at which back-order is due at the warehouse	s9(6) numeric packed

FIGURE 10-5   Format of INVENTORY FILE.

```
CRC420                  CONTROL TOTALS                99/99/99


DATE COMPILED              99/99/99


INPUT RECORDS                                      999,999,999

OUTPUT RECORDS                                     999,999,999

OUTPUT RECORDS TO CRC430                           999,999,999

OUTPUT RECORDS TO CRC440                           999,999,999

OUTPUT RECORDS TO CRC450 .                         999,999,999

OUTPUT RECORDS TO CRC460                           999,999,999

RECORDS REJECTED                                   999,999,999
                                                   -----------

TOTAL RECORDS PROCESSED                            999,999,999
```

FIGURE 10-6   Control totals for CRC420.

10.2.3   Estimate of Auxiliary Storage

HPC has come up with a low bound and a high bound of the number of bytes for each file. Figure 10-11 shows the actual computations leading to these estimates. The following table summarizes these findings:

	Low Bound	High Bound
Master File	68,500,000	280,000,000
Inventory File	250,000	250,000
Sales File	2,100,000	2,100,000
Customer File	480,000	800,000
Totals	71,330,000	283,150,000

The space estimate for all the files are in the range 71,330,000 to 283,150,000 bytes. HPC will be using a 3350 disk drive, which has 30 tracks per cylinder, 19,000 bytes per track, and 556 cylinders, equaling 316,920,000 bytes per disk drive. In order to account for 100,000 bytes for test work space and 350,000 bytes for transaction file space, HPC will need one disk drive.

10.2.4   Data Communication Network

Toy World, Inc. operates three branch stores at Philadelphia, Scranton, and Wilkes-Barre and has its headquarters in Harrisburg (see Section 4.2). HPC

```
CRC430                    CONTROL TOTALS                    99/99/99


DATE COMPILED            99/99/99

INPUT RECORDS                                           999,999,999

MASTER FILE RECORDS UPDATED                             999,999,999

RECORDS REJECTED                                        999,999,999
                                                        -----------

TOTAL RECORDS PROCESSED                                 999,999,999
```
FIGURE 10-7 Control totals for CRC430.

```
CRC440                    CONTROL TOTALS                    99/99/99


DATE COMPILED            99/99/99

INPUT RECORDS                                           999,999,999

SALES FILE RECORDS UPDATE                               999,999,999

RECORDS REJECTED                                        999,999,999
                                                        -----------

TOTAL RECORDS PROCESS                                   999,999,999
```
FIGURE 10-8 Control totals for CRC440.

```
CRC450                    CONTROL TOTALS                    99/99/99


DATE COMPILED            99/99/99


INPUT RECORDS                                           999,999,999

CUSTOMER FILE RECORDS UPDATED                           999,999,999

RECORDS REJECTED                                        999,999,999
                                                        -----------

TOTAL RECORDS PROCESSED                                 999,999,999
```
FIGURE 10-9 Control totals for CRC450.

```
CRC460                 CONTROL TOTALS                  99/99/99


DATE COMPILED          99/99/99

INPUT RECORDS                                      999,999,999

INVENTORY RECORDS UPDATED                          999,999,999

RECORDS REJECTED                                   999,999,999
                                                   -----------

TOTAL RECORDS PROCESSED                            999,999,999
```

FIGURE 10-10   Control totals for CRC460.

recommends that the data processing facility be located in Harrisburg and be connected to the three branch stores via data communication lines. However, the traffic volume between Harrisburg and each branch store is relatively low. Consequently, HPC has designed the following data communication configuration for the complete system:

i. Star type network with the host computer system located in Harrisburg and connected with each branch store via dial-up telephone lines

ii. 300 bps half-duplex asynchronous transmission

iii. Multidrop line configuration, since it is less expensive

Master File
Lowerbound   500,000 records × 137 bytes = 68,500,000 bytes
Higherbound   500,000 records × 560 bytes = 280,000,000 bytes

Inventory File
                                  5,000 records × 50 bytes =   250,000 bytes

Sales File
                                 35,000 records × 60 bytes =   2,100,000 bytes

Customer File
Lower bound   6,000 records × 80 bytes =   480,000 bytes
Higher bound   10,000 records × 80 bytes =   800,000 bytes

Additional Space Needed
Test work space   100,000 bytes
Transaction files space   350,000 bytes

FIGURE 10-11   Space estimates for data files.

HPC recommends that data be transmitted only in bulk and preferably after the peak hours so as to take advantage of reduced transmission rates. Also, since in dial-up connection, a minimum connect time charge is always assessed irrespective of the actual duration of connection, bulk transmission of data will be highly cost effective. Figure 10-12 shows the proposed network configuration.

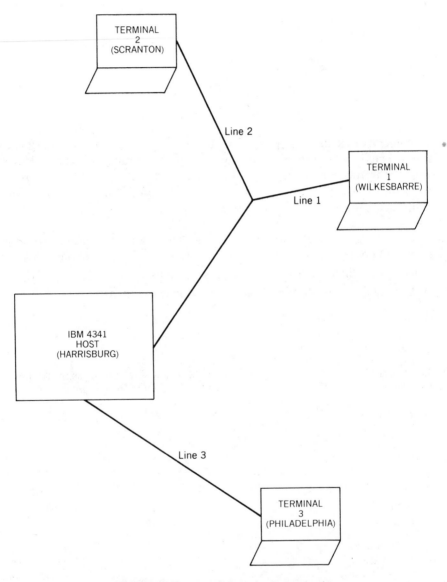

FIGURE 10-12 Data communication network.

QTY	Hardware	ID No.	Specs and Remarks
1	CPU	4341	4 Megabytes Main memory
2	Disk drives	3350	312 Gigabytes
2	Tape drives	3420	1600 bpi
7	Terminals	3270-8	80 columns × 24 lines (giving 1920 display characters)
7	Printers	1403	132 characters per line capable of printing 6 or 8 lines per inch
4	Modems		RS 232-C compatible

FIGURE 10-13 Equipment specification.

10.2.5 Equipment Specification

Figures 10-13 and 10-14 show the equipment specifications and the hardware configuration, respectively, for the complete system. HPC proposes an IBM 4341 computer system for Toy World. Figure 10-15 provides a detailed cost breakdown for the proposed hardware and software. The cost is grouped under two categories: developmental cost and operating cost. The personnel cost is not included in these cost estimates.

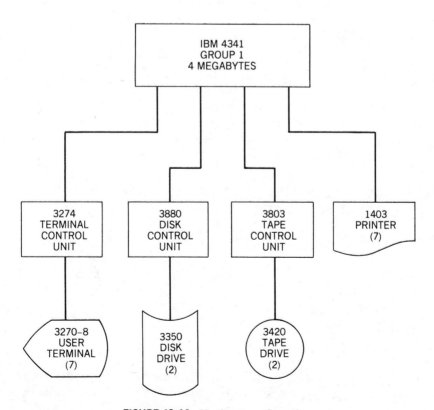

FIGURE 10-14 Hardware configuration.

I. Development Cost	Proposed System	
1. Analysis		$ 16,000
2. Hardware Purchase		
a. IBM 4341		231,716
b. Memory		29,000
c. 3350-A2 disk drive; 2 at $20,540		41,080
d. 3274 terminal control unit		11,500
e. 3880 disk control unit		16,000
f. 3803 tape control unit		14,000
g. 3420 tape drive; 2 at $5,500		11,000
h. 1403 printer; 7 at $800		5,600
Total Hardware Cost		**$359,896**
3. Software Lease		
a. VSE VSAM		$ 720 per year
b. CICS DOS/VS		$ 5,880 per year
c. DOS/VS COBOL		$ 1,884 per year
d. SYNC SORT		$ 3,144 per year
Total Software Lease Cost		**$ 11,628**
4. Software Consultation		$ 3,600
2 at $600 per week × 3 weeks		
Total Development Costs		**$391,124**

FIGURE 10-15 (a) Breakdown of developmental cost.

10.2.6 Personnel

HPC has examined the existing data processing staff at the Harrisburg office of Toy World to determine what additional staff will be needed for the implementation and maintenance of the new system. Figure 10-16 represents their findings, which pertain to the Harrisburg location alone. Since the branch stores will merely send and receive data from Harrisburg but will not do any separate data processing on their own, HPC feels that no additional personnel will be needed there.

10.3 DETAILED SYSTEM DESIGN FOR MASSACHUSETTS EDUCATIONAL FOUNDATION

The financial reporting system for MEF consists of four subsystems (see Section 4.12):

□ Data Entry/Edit/Display
□ Data Validation
□ Database Administration
□ Annual Report Generation

HPC has developed detailed system design specifications for the complete system. To avoid unnecessary details, however, we will only present the

II. Operating Costs
1. Maintenance Costs/Hardware

a. IBM 4341, memory	$ 9,264 per year
b. 3350-A2 disk drive (2)	$ 2,400 per year
c. 3274 terminal control unit	$ 1,200 per year
d. 3880 disk control unit	$ 1,200 per year
e. 3803 tape control unit	$ 2,400 per year
f. 3420 tape drive (2)	$ 1,200 per year
g. 1403 printer (7)	$ 1,800 per year
Total Maintenance Costs/Hardware	**$ 19,464**
2. Utilities	$ 6,000
3. Supplies	$ 4,800
Total Operating Costs	**$ 30,264**

FIGURE 10-15   (b) breakdown of operating cost.

Existing Staff

1. Four keypunch operators to be trained as data entry clerks in the following manner:
   a. Two sales order entry clerks; on-going cost per clerk per year $15,000
   b. Two customer service entry clerks; on-going cost per clerk per year $15,000
2. Computer operator to be trained on IBM 4341; on-going cost $20,000 per year
3. Seven-member systems and programming staff to develop software for new inventory control system:
   a. Manager of programming; on-going cost $42,000 per year
   b. Systems programmer; on-going cost $35,000 per year
   c. Three application programmers; on-going cost $25,000 per year
   d. Maintenance programmer; on-going cost $25,000 per year
   e. Program librarian/documentation; on-going cost $16,000 per year

Support Staff to be Hired

Two software consultants to train and work with the existing staff; three weeks employment; $600 per week per consultant

Staff Reduction

Two shifts of six clerks will be phased out through attrition over a period of 6 months. The following jobs will be phased out:

a. Five clerks at $10,000	$ 50,000	
b. Four clerks at $11,100	$ 44,400	
c. Three clerks at $9,867	$ 29,601	
Total Cost Savings	**$124,001**	

FIGURE 10-16   Personnel cost estimates.

system design specifications for the subsystem, Database Administration. Since Toy World's system does not have a database component, this will be an occasion to explore that area.

Figures 8-25 and 8-26 showed that the two programs, Update Form Database Program and Ad Hoc Query Program, together handle the Database Administration subsystem and interface with the Form Database.

10.3.1 Database Design

An authorized user with the appropriate level of clearance accesses the financial reporting system via Executive Module and is greeted with the AD-MIN version of the main menu (see Fig. 8-33). By selecting the third option, Database Administration, the user views the Database Administration Menu (see Fig. 10-17), which provides access to the Form Database.

The Form Database is a relational database containing the following relations:

   i. MEF form XX
  ii. Database status
 iii. Validation error

Each of the 25 input forms used by MEF (MEF-01, MEF-02, etc.) is stored in a separate relation called MEF Form XX, where XX ranges from 01 to 25. Each relation contains the data entered in each cell of the form, the grantee ID, and the fiscal year. The grantee ID, the fiscal year, and the cell

```
                MASSACHUSETTS EDUCATIONAL FOUNDATION


                    DATABASE ADMINISTRATION MENU



        The following options are available:




                    1.  Create/Update Form Database

                    2.  Generate Query Reports

                    3.  Help

                    4.  Exit


        Select an option and type the corresponding number. Then press
        RETURN.
```

FIGURE 10-17 Database administration menu.

locations (i.e., line and column numbers) are keyed for quick retrieval of data.

The Database Status relation contains the grantee ID, fiscal year, MEF form number, validation status flag, and date/time of the last update of the MEF form. At present, the validation status flag accepts any one of the three values:

00 Not validated
01 Processed by validation routines and has at least one failure
02 Processed as clean by validation routines

Finally, the Validation Error relation contains the grantee ID, fiscal year, MEF form number, offending cells' locations (i.e., line and column numbers of all cells that caused a failure of validation), and short error descriptions.

By selecting option 1 in Figure 10-17, the user can create a new MEF form relation or update an existing MEF form relation. The second option in Figure 10-17 allows the user to access one or more relations in the Form Database and manipulate them using ORACLE (the DBMS to be used by the system) commands. Thus, the user can generate ad hoc query reports.

The Form Database provides a permanent repository of data available in the MEF forms. The Database Administration subsystem interfaces with two other subsystems, namely, Data Entry/Edit/Display and Data Validation. The Database Status and the Validation Error relations contain vital information pertaining to these interfaces.

10.3.2 Equipment Specification

The operations of MEF are centralized in its head office in Boston. All the forms completed by its grantees are mailed to Boston. Hence HPC finds no need to design a data communication network. The following hardware and software recommendations are made:

i. VAX 11/780 computer with 4.0 MB memory and VMS operating system
ii. ORACLE relational DBMS with COBOL interface
iii. Structured COBOL programming language
iv. Two Winchester disk drives with 133 MB per drive
v. One DEC TWE-16 tape drive for backup and archival operations
vi. Three DEC Writer III printing terminals
vii. Two high-speed line printers
viii. Ten VT-180 CRT terminals
ix. FMS software to generate menus and forms

Figure 10-18 shows the hardware configuration for the system.

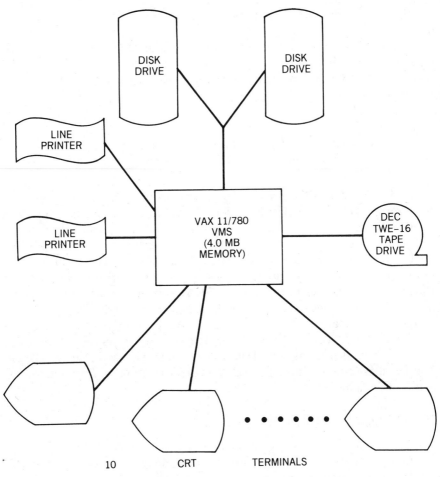

FIGURE 10-18 Equipment configuration for MEF.

10.4 SUMMARY

This chapter completed the detailed system design phase for the two case studies—Toy World and Massachusetts Educational Foundation.

For Toy World, the chapter explored the order processing subsystem built around process 4, i.e., process customer requested changes. The rest of the system can be developed in a similar manner. The following items were included as part of the detailed system design specification:

a. System flowchart of the selected process
b. File design specifications for the four files used in the complete system
c. Estimate of auxiliary storage space

d. Data communication network

e. Equipment specification

f. Personnel estimates

For Massachusetts Educational Foundation, the chapter explored one subsystem, Database Administration. A relational database environment was recommended, and the subsystem was accessed via a menu-driven Executive Module. An overview of the database design was provided. Finally, the chapter gave a list of hardware and software needed for the system.

10.5 KEY WORDS

The following key words are used in this chapter:

auxiliary storage	file access
data communication network	file organization
detailed system design	relational database

REVIEW QUESTIONS

1. Develop the inventory control subsystem and the sales management subsystem for Toy World.
2. Using the space estimate methodology developed in Section 9.7.2, compute the number of tracks needed for storing the four files used by Toy World's system. What blocking factor do you use?
3. Develop the data entry/edit/display subsystem for MEF. How do you want to interface this subsystem with the database administration subsystem? Give all the details of your development.
4. Assume that you have selected option 2 in Figure 10-17. Work out a complete operational scenario describing the generation of an ad hoc query report.
5. HPC has recommended ORACLE as the DBMS for MEF's proposed system. Using references such as DATAPRO to gather information on ORACLE, specify (with reasons) the options that ORACLE should have in order to handle all the requirements of the proposed system.
6. How would you assess the advantages of choosing a block structured language such as C or Pascal instead of structured COBOL to implement the MEF system?
7. Estimate the auxiliary storage needed by the Form Database. Specify all the assumptions you make to come up with your estimate.
8. Comment on the number and type of personnel needed to implement and maintain the system for MEF.

STRUCTURED IMPLEMENTATION

Part IV consists of Chapters 11 and 12. Chapter 11 discusses the theory of system implementation, maintenance, and evaluation, which is the final phase of the system life cycle. The chapter first describes the detailed program level design, which consists of input–output tables, structure charts, HIPO charts, program flowcharts, and pseudocodes. Then follow the system implementation topics: system documentation, user training, system backup and recovery plan, and system conversion. Finally, system maintenance involves both preventive maintenance as well as rescue maintenance. Chapter 12 completes the two case studies through the system implementation phase.

11

System Implementation, Maintenance, and Evaluation: Theory

11.1 SCOPE OF THE CHAPTER

System implementation is the last phase of the system development process. Its goal is to convert the proposed system from paper design into electronics. As such, the contents of this chapter can be divided into three main components:

a. Program level design, which includes
- ☐ Input/output table
- ☐ Structure chart
- ☐ HIPO chart
- ☐ Program flowchart
- ☐ Pseudocode

b. System implementation, which includes
- ☐ Structured programming
- ☐ Testing and debugging techniques
- ☐ Preparation of physical site
- ☐ User training
- ☐ System conversion

☐ System backup plan and audit trail
☐ System recovery plan
☐ Documentation and user manuals
☐ Project management

C. System maintenance and evaluation

The program level design is essentially the counterpart of system level design and is often included in the detailed system design phase. It addresses the detailed design issues at the individual program level rather than at the subsystem level. System implementation involves the actual coding, testing, and debugging of programs until they run error-free. The rest of this phase involves adequate support to run the operational system. As such, the support includes adequate documentation, user training, backup and recovery plans, and an overview of project management principles. System maintenance and evaluation become pertinent after the system is operational.

11.2   INPUT/OUTPUT TABLE

An *input/output table* provides a link between the system level design and the program level design. The table consists of two columns, Input and Output. For each program appearing in the system flowchart of a given subsystem, the table lists all inputs and all outputs. The input entries can be source documents, data files, or information passed from an earlier program; the output entries can be reports, screen displays, data files, or information passed to other programs. The entries of the input/output table help the system team prepare the structure chart or the hierarchy chart of each program.

As an example, consider the system flowchart in Figure 7-11. It has three programs:

☐ Data Entry and Edit Program
☐ File Update Program
☐ Report Generation Program

The input/output table for the flowchart is as follows:

Program: Data Entry and Edit

Input	Output
Data from CRT	Error Report
Transaction File Record	Data to File Update Program
	Transaction File

Program: File Update

Input	Output
Data from Data Entry and Edit	Reject Report
Master File Record	Data to Report Generation Program
	Master File

Program: Report Generation

Input	Output
Data from File Update Program	Management Reports
Master File Record	Master File

11.3 STRUCTURE CHART

A *structure chart* gives the programmer a visual picture of the overall architecture of the program. The basic building block of a program is a *module*. The structure chart portrays this architecture by arranging the modules in a hierarchical manner. The relationships of the modules to each other are defined, and the interfaces between modules, along with the method of control, are identified. Thus, the structure chart includes a complete description of the system with respect to modules, data, their interrelationship, and control.

Submodules of a module are identified at a level below the module, with the data sent from the module to the submodule and from the submodule back to the module identified along the connecting line. Arrows indicate the direction of flow. The data passed between modules may be used either as information to modules or as a control flag to the module. The terminals of the arrow indicate the type of flow, whether data or control.

11.4 PRINCIPAL CHARACTERISTICS OF A STRUCTURE CHART

In order to evaluate the design of a structure chart, several guidelines are used. We discuss five criteria that can be used to examine the structure for determining the quality of a structure chart:

a. Coupling. Coupling is a measure of the relationships that exist between modules. We measure coupling by the number of connections as well as by the type of connection and the type of information communicated in that connection. Thus, a low measure of coupling is desirable since it indicates that the individual modules are fairly self-contained.

b. **Cohesiveness or Cohesion.** Cohesiveness is a measure of the type of relationships that exist between elements in the same module; that is, the binding between statements. The stronger the relationship, the more likely the module can and should be viewed as a single unit. Thus, high cohesiveness is desired. Functional cohesiveness is by far the strongest form of cohesiveness. It implies that all the elements of the module perform a single function.

c. **Scope of Effect/Scope of Control.** The scope of control of a module consists of that module plus all the modules ultimately subordinate to it; that is, the scope of control encompasses all nodes in all branches emanating from the module. The scope of effect of a module consists of all modules affected by a decision of that module. The system is simpler if the scope of effect is contained within the scope of control. The failure to conform to this principle results in an increased number of control flags that must be passed between modules. This has been found to result in high coupling and increased complexity.

   Figure 11-1 represents a structure chart in which the scope of effect is *not* contained within the scope of control. Here the scope of control of module B consists of B, D, and E, whereas the scope of effect of B consists of B and C.

d. **Size.** No strict rules exist that dictate the ideal size (i.e., number of statements) of a module. The size depends on the programming language used. For instance, COBOL uses more statements than FORTRAN to implement the same function. Normally, 30 to 50 statements are regarded as an ideal size. Size is also closely related to human comprehension.

e. **Abstraction.** During initial stages of design, the structure has a very important characteristic—it is independent of actual implementation constraints and is expressed in abstract form. Abstract items are uncommitted to any hardware or data configuration.

11.5 PROCEDURE FOR DESIGNING A STRUCTURE CHART

A program consists of three main components:

   ☐ Input
   ☐ Processing
   ☐ Output

It receives input data, processes it, and then outputs the result. Accordingly, a structure chart of a program consists of three main branches:

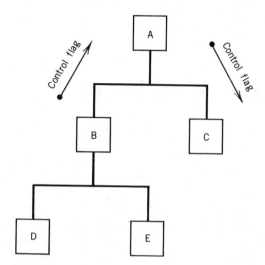

FIGURE 11-1 Scope of effect versus scope of control.

a. **Afferent branch**, which receives the input and formats it for processing
b. **Transform branch**, which does the actual processing
c. **Efferent branch**, which formats the data for output

For the entire chart as well as within each branch of a chart, nodes at the top or intermediate levels serve solely as control functions to their subordinates. The actual processing is done at the terminal nodes. The intermediate nodes define why the terminal nodes are required and as such provide continuity for the entire program.

In a structure chart it is mandatory that only the terminal nodes interface with hardware either by performing an input/output operation or by doing some kind of processing such as testing a decision condition, working in a loop, or executing algorithms. All nonterminal nodes merely receive data from their subordinates and pass data on to other nodes. Also, there cannot be any communication between any two branches unless the communication passes through the topmost node, which is called the *control module.*

The design of structure charts follows a principle called the law of conservation of variables. Jensen and Tonies have described the law as follows ([4], p. 181):

It is based on the premise that data is only *created* external to the software system. Within the software system, data is transformed until it is of the form required for output. The software system itself, however, never creates new information; it creates only changed information. Therefore, the data to be used by a branch is passed down the branch until it reaches a terminus node—the processing module where the transformation is performed. The output parameter is transferred up the structure from the point where it was created to the point where it was used.

11.6 SYMBOLS USED IN A STRUCTURE CHART

The structure chart begins with the control module, shown as a rectangular box at the top level.

Subordinate modules are shown below the superordinates, usually in left to right order, as they appear lexically. References between modules are represented with arrows.

Ordinary arguments and data elements are represented by a small arrow with circles on its tail. Control data are represented by a small arrow with a dot on its tail. (Control data are used to influence execution. An example is flags, which represent the outcome of decisions and "tell" the program what to do.) The direction of the arrow indicates whether the data are input to or output from a node.

In addition to structure, certain procedural information is shown on the structure chart.

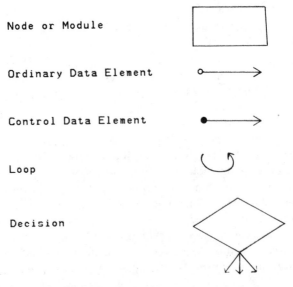

FIGURE 11-2 Symbols in a structure chart.

Loops are indicated by a looping arrow that encompasses the tails of the arrows for the connections used in the loop. Loops can be nested.

Conditional executions are represented by diamond-shaped figures. The number of arrows coming out of a diamond corresponds to the number of outcomes of a decision. When several calls are made as the result of one outcome, the outcome arrows start at the same point in the diamond. Procedural annotation can be used with any type of connection.

Figure 11-2 shows the symbols used in a structure chart.

11.7 EXAMPLE OF A STRUCTURE CHART

We now illustrate the process of designing a structure chart from a data flow diagram using a reorder processing subsystem of an inventory control system. The subsystem ensures that the company can maintain inventory as low as possible while still provide an adequate service level for its customers. The design principle of the subsystem is as follows:

For each item in the inventory, the company determines an optimal reorder level by using an operations research model of inventory management. On a regular basis (daily, weekly, etc.) the quantity on hand for an item is compared with the optimal reorder level for that item. If the former is less than or equal to the latter, a purchase requisition request is sent to the Purchasing Department so as to replenish the stock of that item; otherwise, no action is taken. This policy ensures that under normal demand the company will have adequate stock to fill customer's orders but not tie up too much capital in the form of unused inventory.

Figure 11-3 shows the data flow diagram for the subsystem; Figure 11-4 is a partial data dictionary for this subsystem; and Figure 11-5 gives the structure chart.

11.8 HIPO CHARTS

HIPO stands for *H*ierarchy and *I*nput *P*rocessing *O*utput. This technique shows the hierarchical structure of the modules constituting a program, as well as other information on each module. The HIPO charts consist of two types of charts:

a. One hierarchy chart that shows the top-down structure of a program by decomposing it into the component modules

b. An *I*nput *P*rocess *O*utput (IPO) chart for each module in (a), which describes the inputs to, the outputs from, and the logic of the process performed by the module. The data dictionary created earlier supplies the necessary information to prepare an IPO chart.

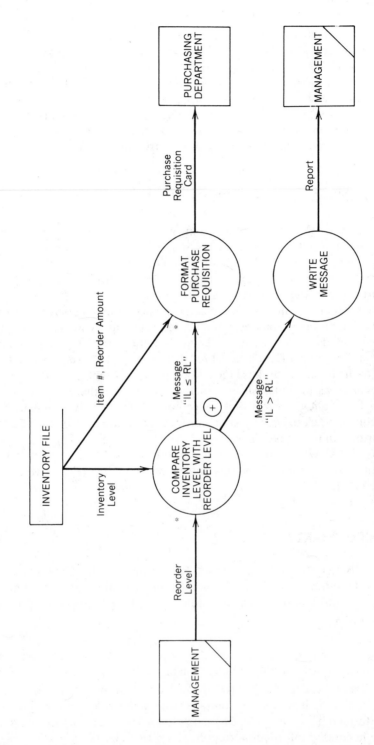

FIGURE 11-3 Data flow diagram of reorder system.

```
SOURCE:          MANAGEMENT

DESTINATIONS:    MANAGEMENT, PURCHASING DEPT.

PROCESS LOGIC:   Compare optimal reorder level with current

                 inventory level to determine need to order.

                 Initiate such order or issue report to the

                 contrary.
```

FIGURE 11-4 Partial data dictionary for the reorder subsystem.

Collectively, a set of HIPO charts performs the same function as a single structure chart, since both of them describe the program logic graphically. For very complex programs, the structure chart may look very cluttered due to a large number of arrows with the names of data elements written beside them. In such cases, a hierarchy chart is preferable, because its appearance is identical with a structure chart except for the arrows representing data elements, the names appearing against these arrows, and other conventions such as looping arrows and decision boxes. A set of IPO charts accompanying the hierarchy chart then conveys the complete logic of the program.

Davis ([3], p. 326) has summarized the advantages of the HIPO charts:

> The HIPO documentation serves a number of purposes. By using it, designers can evaluate and refine a design, and correct flaws prior to implementation. Given the graphic nature of HIPO, users and managers can easily follow a program's structure. Finally, programmers can use the hierarchy and IPO charts as they write, maintain, or modify the program.

11.9   EXAMPLE OF HIPO CHARTS

Our example of a HIPO chart involves a patient invoicing system used by a small group of physicians forming an associates group. At the time a patient visits a physician, the latter manually fills out a Diagnosis and Treatment Services form describing the services provided and drugs administered, if any. The program, Generate Charges, accepts the input data from this form, calculates the total charge for which the patient is to be billed, and writes the output in the Patient Data File. Each record in the Patient Data File has a fixed part containing data such as the patient's name, address, and account number and a variable part containing data pertaining to each visit of the patient such as total charge, diagnosis codes and descriptions, and drugs given to the patient. The program, Generate Charges, writes the output in the variable part of the patient's record.

Figures 11-6 and 11-7 represent, respectively, the structure chart and the

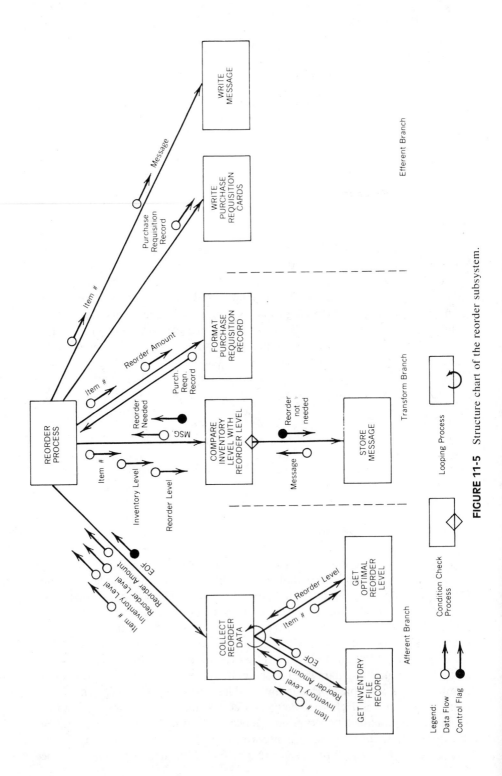

FIGURE 11-5 Structure chart of the reorder subsystem.

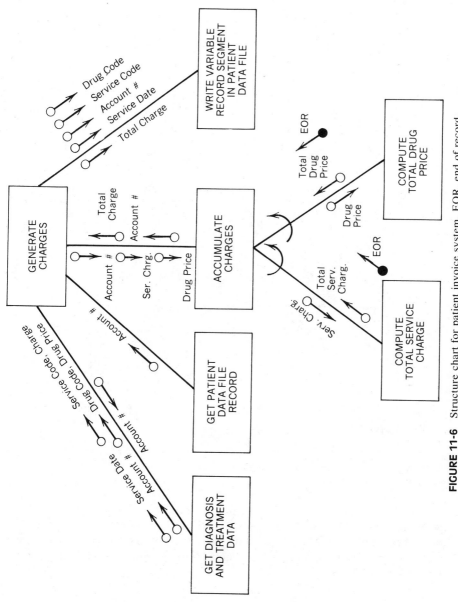

FIGURE 11-6 Structure chart for patient invoice system. EOR, end of record.

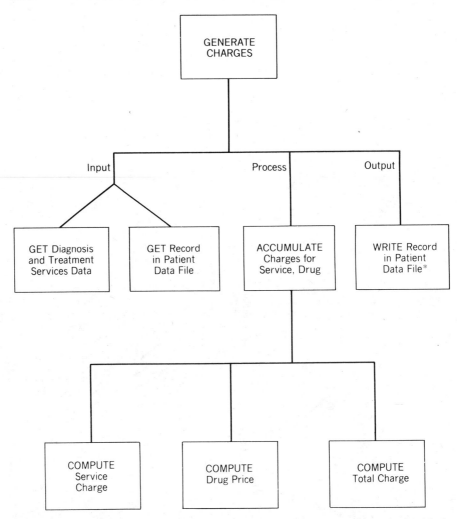

FIGURE 11-7   Hierarchy chart for patient invoice system. *This process adds a new data block in the variable segment.

hierarchy chart of the program. Note that the only difference between these two figures is that Figure 11-6 contains the data flow arrows, looping arrows, and decision boxes. Figures 11-8 through 11-10 contain three IPO charts for three modules.

11.10 PROGRAM FLOWCHARTS

A program flowchart can be regarded as an alternative to an IPO chart. The IPO chart describes the processing logic, input, and output of a module in

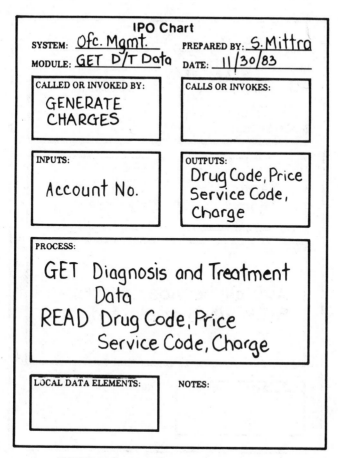

FIGURE 11-8 IPO chart for an input module.

a structured form using structured English. A program flowchart provides the same information in graphic form. By convention, the flow of logic goes from top to bottom and from left to right. Figure 11-11 is the program flowchart for the IPO chart given in Figure 11-9.

The architecture of a complex program can be described in three ways:

a. A single structure chart
b. A single hierarchy chart and a set of IPO charts, one for each module in the program
c. A single hierarchy chart and a set of program flowcharts, one for each module in the program

As a rule of thumb, if a program has more than 5 levels or has more than 15 modules, the structure chart becomes very cluttered so that option (a) is

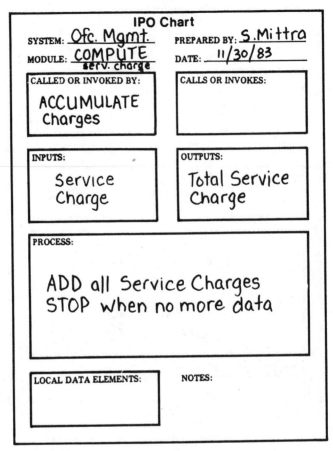

FIGURE 11-9   IPO chart for a process module.

no longer suitable. In that case, either (b) or (c) should be selected to design the program.

11.1 PSEUDOCODE

Pseudocode is like real code, and its structure is usually based on the structure of a real programming language. It is regarded as an intermediate step between a structure chart or HIPO charts and actual codes. Pseudocode often appears in the form of structured English. However, it avoids details such as opening and closing files, initializing counters, and setting up flags. Once pseudocodes are generated from structure or HIPO charts, the writing of actual code becomes routine.

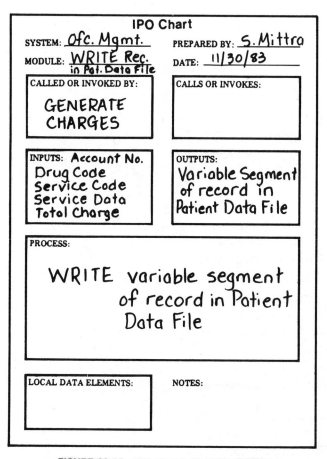

FIGURE 11-10 IPO chart for an output module.

The main difficulty with pseudocodes is that no universal standards exist to define them. As a result, pseudocode for FORTRAN is very different from that for COBOL or Pascal, say. Hence, the question arises; If a pseudocode is so close to an actual programming language, why should one waste time in writing pseudocodes instead of writing the actual programs? In addition, if the pseudocodes are generated by analysts, then the programmers often resent these codes since the latter do not allow flexibility of programming. It appears that if the processing logic of each module of each program is available in the form of an IPO chart or structured English, then pseudocodes are not needed.

Figure 11-12 gives a detailed example of pseudocodes for updating an inventory master file. It is written in the style of a block-structured language such as Pascal or PL/1.

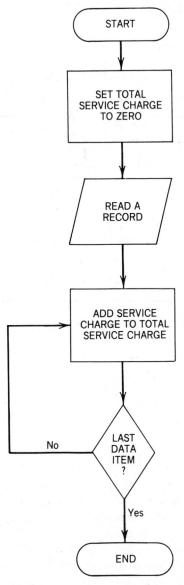

FIGURE 11-11 Sample program flowchart for COMPUTE total service charge.

11.12 STRUCTURED PROGRAMMING

Structured programming can be visualized as the application of basic program description methods whereby a program is written in hierarchical structure. It presupposes a top-down approach (as opposed to a bottom-up approach) and breaks up a large complex program into small manageable modules. Program stubs, which are empty modules, are used to simulate functions of modules not yet developed. Coding starts from the top module and proceeds downward.

In 1966, C. Böhm and G. Jacopini ("Flow Diagrams, Turing Machines, and Languages with Only Two Formation Rules," Communications of the ACM, vol. 9, pp. 366–371) first introduced the mathematical basis of structured programming in their paper. They proved that any program can always be written using three basic control structures:

a. **Sequence.** Program instructions are executed in the order in which they appear.

b. **Decision.** The sequential order of statement execution is broken, and control branches to a nonsequential instruction on the basis of a decision criterion. This is called *conditional branching*.

c. **Repetition.** A given sequence of statements is executed a finite number of times depending on the outcome of an exit criterion. After each iteration of the statements, the exit criterion is evaluated to determine if the iteration should be continued. This is the typical *looping* mechanism.

Figure 11-13 shows these three basic control structures.

Depending on the programming language used, sequence is implemented via computational formulas and LET statements; decision is implemented via IF-THEN-ELSE statement; and repetition is implemented via DO WHILE, FOR NEXT, REPEAT UNTIL, etc., statements. As is seen in Figure 11-13, each control structure has a single entry and a single exit.

The use of the single entry/exit structure in organizing logical structures allows the clustering of program functions into single entry/exit modules or subsystems. Assigning program fuctions to specific modules allows the modules of a complex program to be independently programmed by separate programmers. Because each module has only one point of entry and one point of exit, the interfacing of these modules into one program or system is simplified. Since each module's function is clearly defined before it is programmed, the modules fit together to form a complete program.

Besides allowing program functions to be programmed independently, modularization simplifies program testing, debugging, and modification. Since program functions are defined in specific modules, they can usually be tested, debugged, and modified as separate modules of program code. This reduces

Update Inventory Master File

Functions

1. Nightly sequential update of Inventory Master File
2. Report records are created for The Master Listing, Accepted Transactions, Rejected Transactions Report, Accepted Field changes, and Rejected Field Change Report.

Input/Output

Input consists of
FINAL EDITED TRANSACTION FILE
INPUT MASTER FILE
Output consists of
OUTPUT MASTER FILE
REPORT FILE

Subroutines

HOUSEKEEPING: Clear historic fields on Master File dependent on dates
GET-MASTER: Reads Master File
GET-TRANS: Reads Transaction File
PUT-MASTER: Writes updated Master Records
RECEIPT ROUTINE: Processes additions to inventory
ISSUE ROUTINE: Processes items withdrawn from inventory
FIELDS CHANGE ROUTINE: Processes field changes to Master File;
adds and deletes items to inventory
REORDER ANALYSIS: Determines if item needs to be reordered and
generates reorder form
CREATE-MASTER-LIST: Generates report records for Master Listing
ACCEPTED-TRANS-RPT: Generates report records for Accepted Transaction Report
REJECTED-TRANS-RPT: Generates report record for Rejected Transaction Report
ACCEPTED-FIELD-CHG-RPT: Generates report records for
Accepted Field Change Report
REJECTED-FIELD-CHG-RPT: Generates report records for
Rejected Field Change Report

Pseudocodes:

```
UPDATE (MAIN ROUTINE)
    CALL HOUSEKEEPING
    CALL GET-MASTER /* read first Master/*
    CALL GET-TRANS /* read first Trans/*
```

```
DO WHILE (there are more records)
    When (ITEM# TRANS>ITEM# MASTER) DO
        CALL CREATE-MASTER-LIST
        CALL REORDER-PROCESS
        CALL PUT-MASTER
        CALL GET-MASTER
    END
    When (ITEM# TRANS=ITEM# MASTER) DO
        When (TRANS-CODE-TR<50)
            CALL FIELD CHANGE ROUTINE
        When (TRANS-CODE-TR = 50)
            CALL RECEIPT ROUTINE
        When (TRANS-CODE-TR = 60)
            CALL ISSUE ROUTINE
        Otherwise
            Set Error Code
            Reject Transaction
            CALL GET-TRANS
    END
    When (ITEM# TRANS<ITEM# MASTER) DO
        If TRANS-CODE-TR = 20 then set up New Account
        ELSE set Error Code and reject Trans
        CALL GET-TRANS
    END
END (DO WHILE)
DO WHILE (there are more records in Master File)
    CALL CREATE-MASTER-LIST
    CALL REORDER-PROCESS
    CALL PUT-MASTER
    CALL GET-MASTER
END (DO WHILE)

DO WHILE (there are more transaction records)
    Set Error Code
    Reject Trans
END (DO WHILE)
```

FIGURE 11-12 Sample pseudocode.

Update Inventory Master File

```
HOUSEKEEPING ROUTINE
  If (last day of week) clear WEEKLY-SALES-MF
  If (last day of month) clear MONTHLY-SALES-MF
  If (last day of year) clear YEAR-TO-DATE-SALES-MF
  If (first day of year) then
    LAST-YEAR-SALES = YEAR-TO-DATE-SALES-MF
  END

RECEIPT ROUTINE
  Add SHIPPING-QTY-TR to QTY-ON-HAND-MF
  Set DATE-LAST-RECEIPT to today's date
  Subtract SHIPPING-QTY-TR from QTY-ON-ORDER-MF
  CALL ACCEPTED-TRANS-RPT
  END

ISSUE ROUTINE
  Subtract SALES-QTY-TR from QTY-ON-HAND-MF
  Set DATE-LAST-ACTIVITY-MF to today's date
  Add SALES-QTY-TR to DAILY-SALES-MF, to WEEKLY-SALES-MF, and to MONTHLY-
    SALES-MF
  CALL ACCEPTED-TRANS-RPT
  END

FIELD CHANGE ROUTINE
  When (TRANS-CODE-TR = '10')
    CALL NEW-ITEM
  When (TRANS-CODE-TR = '20')
    CALL DELETE-ITEM
  When (FIELD#-TR = 2) DO
    Save (old item description)
    ITEM-DESCRIPTION-MF = FIELD-VALUE-MF
    CALL ACCEPTED-FIELD-CHG-RPT
  END
```

```
When (FIELD # -TR = 3) DO
    If FIELD-VALUE-MF not numeric then
        CALL REJECTED-FIELD-CHG-RPT
    ELSE save old UNIT-COST-MF
        UNIT-COST-MF = FIELD-VALUE-MF
END
Etc. (for every field on Master File)
END

REORDER ANALYSIS
If QTY-ON-HAND-MF<REORDER-POINT then
    issue Purchase Requisition Record
    Format Purchase Requisition Card
END
```

FIGURE 11-12 *continued*

273

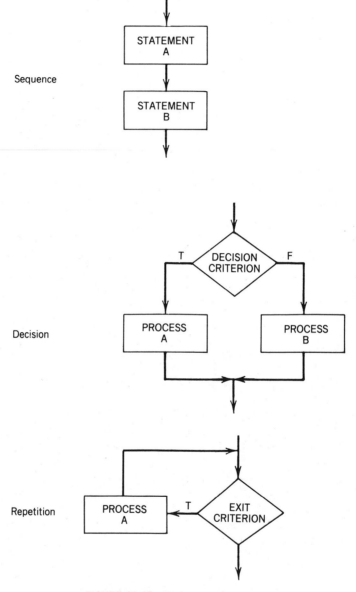

FIGURE 11-13   Basic control structures.

confusion caused by program statements branching into or out of modules at points other than where modules are designed to be entered or exited.

The execution of a computer program terminates after a finite number of steps if and only if the repetition control structure has a finite exit criterion. For example, a repetition characterized by the statement

```
WHILE P DO S
```

implies that the iteration will continue as long as P is true. Hence a necessary condition for preventing an infinite loop is that S must change the value of some variable in the condition P. Prather ([6], p. 128) has given the following conditions to assure finiteness of a loop:

If there exists an integer-valued function F of some of the variables of the exit criterion such that

i. If P is true then $F > 0$

ii. $F_{n+1} < F_n$, where F_i is the value of F in the ith iteration of the loop,

then the loop is finite.

Here we note that by (ii), the function F is a strictly descending integer-valued function. Hence eventually F must become nonpositive. At that time, by (i), P becomes false and hence the loop terminates.

None of the three basic control structures used in a structured program includes GOTO, which is an unconditional branch statement. This situation has sometimes led to the misconception that a GOTO-less program is a structured program. A note from E. Dijkstra ("GOTO Statement Considered Harmful," Communications of ACM, 1968, vol. 11, pp. 147–148) has further compounded the confusion. Dijkstra proved statistically that for programs written in an ALGOL-like language the presence of GOTO statements increases the number of errors and execution time. Accordingly, some people believe that by merely eliminating GOTOs from a program, the program becomes structured. Nothing can be farther from the truth.

In conclusion, a structured program is written using the three control structures

☐ Sequence
☐ Decision
☐ Repetition

and avoiding GOTO as far as practicable. The role of GOTO is played by subroutine calls such as CALL in FORTRAN, PERFORM in COBOL, and GOSUB in BASIC. However, occasional appearance of GOTOs in a well-designed structured program does not make it unstructured.

11.13 TESTING AND DEBUGGING TECHNIQUES

Testing involves the verification of the basic logic of each program module and the verification that all the modules together work properly. Any program of moderate to high complexity contains many possible branching instructions. As a result, an attempt to test all possible paths in the total program logic often leads to combinatorial explosion. Consequently, instead of being exhaustive in the testing of modules, we normally follow the most likely paths to ensure that the modules work properly.

In structured programming, the coding starts from the top module and proceeds downward (see Section 11.12). Accordingly, during the testing of the nonterminal upper modules that perform control functions only, dummy data are passed to simulate the operation of the terminal modules. This process is called *stub testing*, and the dummy operations are called *stubs*. Eventually, when all the modules have been coded and tested individually, it is necessary to test the complete program by using a set of data that can test all the modules in the program. This process is called *string test* or *link test*. A successful string test that checks all the major logical paths in the program gives substantial validity to the program, although it does not guarantee that the program is absolutely error-free.

During the string test, both valid and invalid data should be used in order to check the various branches. For example, in testing edit programs, we should use

a. Valid data to ensure that they pass the edit checks
b. Invalid data to verify that they are sent to appropriate error reports

Debugging is done during the execution and testing of programs. Syntactic errors are tracked during compilation time; run-time errors are captured at the execution time. If, however, no error messages appear when the program is run and, at the same time, the desired output is also not produced, the program may have some flaw in logic. A standard debugging technique to address this situation is to insert plenty of "print" statements (e.g., PRINT in BASIC, WRITELN in PASCAL, DISPLAY in COBOL) in the main body of the program. For example, if module X has a statement,

```
PRINT "MODULE X IS RUN",
```

which does not appear in the output, it becomes clear that the program logic does not invoke module x. The programmer can then take corrective steps to alter the logic of the program.

Automated debuggers are now available for many programming languages. Debugging versions of compilers, which facilitate traces of program logic and the inclusion of special debugging information, are also available. The CHECK function in PL/1, for example, when preceding a procedure, prints all program variables when they change values and prints the names of called subroutines.

11.14 PREPARATION OF A PHYSICAL SITE

It is necessary to have a detailed plan for preparing the area in which the computer and its peripherals will be located. Such a plan must be implemented before the computer arrives. This plan really belongs to engineering rather than to system development. However, since this is an essential part of the total implementation strategy, the system team should address the issue. The main items of the site preparation plan are listed below ([3], pp. 164–166):

1. **Power Supply.** Consistent power supply is crucial for computer performance. When power is lost, or even when brownout occurs due to lower than normal voltage, precision is sacrificed. Due to power outage, valuable data may be lost. In order to remedy this situation, the computer installation can use an *uninterruptible power supply* (UPS) that will provide supplemental power to drive the system at a constant voltage.

2. **Power Surges.** A power surge can occur when lightning strikes a line or when heavy equipment located somewhere on the line is suddenly turned off or on. This can cause minor to substantial damage to the hardware. In order to avoid this potential damage, an inexpensive *surge arrestor* should be placed somewhere between the power supply and the computer.

3. **Air Conditioning.** Computer equipment does not function well in high temperature and high humidity. In addition, the equipment itself generates a lot of heat. Consequently, a good air conditioning system is vital for the proper functioning of computers.

4. **Underground Cables.** A mini or a mainframe system has several large parts that must be connected for proper functioning. Large cables are used to connect these parts. Accordingly, the floor of the computer room must be raised in order to pass these cables underneath.

5. **Adequate Storage Areas.** Storage areas are needed to keep supplies such as printer paper, disk packs, diskettes, and tapes. A separate vault should be planned as the tape library.

6. **Furniture.** Furniture includes desks, tables, work surfaces, storage cabinets, book stands, and terminal pedestals. If the existing furniture is not adequate, new pieces must be purchased and installed before the equipment arrives.

11.15 USER TRAINING

Intensive user training is essential for successful implementation of a system. Since users are the evaluators of the system in the long run, they must know how the system operates and how they can get the necessary information from the system. It is indeed true to say that satisfied users provide the greatest support to a newly implemented system. However, even a well-designed system can appear intimidating to new users unless they are trained on how to use it. Thus, user training constitutes a good selling feature for a new system.

Usually there are two categories of users:

a. Heavy or regular users

b. Casual users

Heavy users typically consist of clerical personnel who enter data into the system and generate routine reports from the system. Casual users are mostly managers who access the system to generate ad hoc reports or terminal displays. Since the needs of these two groups of users are quite different, groups should receive different types of training. The heavy users should be trained in processing transactions, operating the computer terminals, and possibly preparing data entry sheets to enter input data into the sytem. Casual users should be trained on the format and contents of reports and terminal displays and on the procedures for requesting or generating ad hoc reports.

The quality of user training depends primarily on the instructor. So, appropriate people must be selected to conduct the training. The instructor must be able to look at the system from a user's viewpoint and anticipate typical questions and concerns of users. Since programmers are too close to the system and look at it from a designer's viewpoint, they are normally not good instructors.

Once an instructor is selected, he or she should set up a training schedule and prepare training materials. The materials must be geared toward the type of users. In other words, training materials for heavy users must be different from those for casual users. The actual training session should consist of a brief introductory lecture describing the system (perhaps using viewgraphs) followed by hands-on training, with a terminal for each user. It is preferable to give each user a set of training materials to use during the sessions. There should not be more than ten users per session.

It is often advantageous to have representatives from different departments in the user organization conduct training sessions within their respective departments. Clerical and managerial personnel often feel less intimidated by the new system if it is explained to them by someone from their own department rather than by an outside instructor.

11.16 SYSTEM CONVERSION

System Conversion is the process of changing from the old system to the new system. Conversion requires careful planning to establish the basic approach to be used well in advance of the actual change. Two major planning decisions that establish the basic conversion methodology pertain to the degree of parallel processing to be used and the sequence of conversion activities.

There are two types of system conversion: *parallel* and *discrete*. In a parallel conversion, both the old and the new systems are concurrently processed until the new system stabilizes. This ensures the reliability of the new system before abandoning the old one. In a discrete conversion, the change to a new system terminates the use of the old system. Thus, parallel conversion uses a gradual, or phased-in, approach, whereas discrete conversion is strictly a one-time process.

Parallel conversion can be either *complete* or *partial*. In complete parallel conversion, both the old and the new system run together for a predetermined period of time. During this time, the outputs from both systems are compared. If they agree, the old system is discontinued after the specified time. If discrepancies are noted, the new system is examined and modified to remove the defects and is then run again in parallel with the old system. This process is repeated until the new system is error-free, at which time the old system is discontinued.

In partial parallel conversion, the old system runs in full and the new system is introduced gradually, say, by departments, branches, plants, or other logical breakdowns. Thus, for example, when the new system works satisfactorily in one department, it is implemented in another department. This process continues until it is operational in the entire organization. During this time, the old system is run in parallel to test the validity of the new system and also as a fallback system.

Parallel conversion reduces the risk of implementing a new system. However, it is expensive, since two systems are kept operational at the same time. For example, suppose that an organization is converting from batch to online processing of customer orders. Parallel processing during the implementation requires preparing the transaction documents for keypunching (the old sytem) in addition to entering the transactions through terminals (the new system). Card-punch machines and personnel will still be required as will the computer-processing and report-generating functions of the old system. If the computer that supports the online processing is to replace an old computer, both the new and the old computers are required until conversion is completed and parallel processing is terminated.

In discrete conversion, rigorous testing of the new system is critical before implementation, because the old system is no longer available as a fallback.

It is generally advisable to begin conversion with segments of the system that appear to offer the least difficulty. For example, conversion may start with the best-managed and best-organized functions with the shortest cycles. This enhances the probability of initial successes, which in turn enhances the credibility and momentum of the new system and its implementation.

During the conversion, management should attempt to minimize any additional internal or external requirements on the department undergoing conversion. A department is generally disrupted enough during conversion without having to handle additional disturbances.

A checklist of conversion activities and the sequence should be made to ensure that all bases are covered. Such a document defines responsibilities and lists schedule dates for conversion activities.

The final task of the conversion sequence is to define what will signify the complete implementation of the new system. This definition may vary. Whatever the definition of implementation, it should be agreed upon in advance by organizational members. When the defined implementation has been achieved, the system should be formally approved and accepted, and implementation should be considered completed.

11.17   SYSTEM BACKUP PLAN

A system backup plan consists of a set of preestablished procedures to maintain duplicate copies of all data and program files. This plan is necessary in order to handle situations that can destroy the data on tapes and disks, such as program errors, human errors, hardware malfunctioning, fires, floods, and other natural disasters. Any of these mishaps can put the business of a company in serious jeopardy. A well-designed backup plan helps a company overcome such disasters.

Under the backup plan, all files are copied onto tapes on a regular basis. A sequential file on tape always has a backup in the form of the old file, because whenever the file is updated a separate new file is created using the old file. Random access files on disks are updated using new transaction records, and new data are written on the old data. Consequently, random access files do not have automatic backups. Therefore, on a regular basis, say daily, all the transactions are copied onto tapes, and then perhaps once or twice a week the complete current disk file is copied onto tapes. In order to create a copy of the current disk file, the latest version of the file on tape is processed against the daily transactions that have occurred after the last backup. In this way, it is possible to create a copy of the current version of any random access file by using the backup files on tapes.

Usually, two versions of any master file are kept as backups, often called the three-generation backup method, namely, grandfather-father-son backup strategy.

A backup plan should answer the following questions, as a minimum:

a. What medium will be used to backup?

b. How often should disks be backed up?

c. Where will the backup media be stored?

d. Who will be responsible for the backup?

Davis ([3], p. 169) comments as follows about the backup plan:

Backup is a form of insurance; it must be done in case the unlikely ever happens.

The above discussions pertain to a file processing environment. If a database management system (DBMS) is used, the software itself provides adequate tools to backup data and ensure data integrity. For example, a relational DBMS called LOGIX offers a special feature called *dated relations*. Under this feature, whenever a relation is created, LOGIX automatically creates its backup relation. The backup includes all the attributes of the original relation along with two additional attributes that record the information of when (i.e., date and time) a new value was entered and when it was deleted. As a result, for each relation in the database, a complete transaction log of updates of that relation is automatically created. This transaction log "time stamps" each update of a relation. Using commands such as

```
ROLLBACK   ASOF,   HIST   ASOF
```

it is possible to create a copy of any relation in the database as it existed on a specific day and time as supplied by the user.

11.18   AUDIT TRAILS

An *audit trail* provides the capability to reconstruct a transaction from final output via processing back to input. It thus traces a transaction from its final to initial stage. Establishment of an audit trail for any transaction is a prerequisite for conducting a data processing audit function. The latter is done in an organization in order to ensure that the internal operations are conducted according to the procedures described in the operations manuals. A data processing auditor reviews these manuals and determines if the functions are performed as described there.

There are two types of data processing auditors: internal and external. Internal auditors serve groups within the organization such as executive management, whereas external auditors serve external groups such as stockholders and government agencies. Traditionally, internal auditors have been interested in operating procedures, and external auditors have been interested in rendering an opinion on financial statements. These two areas are not mutually exclusive, because in the final analysis both audit groups are concerned with the integrity and efficiency of the computer-based information system, the output it produces (e.g., financial statements), and its effect upon the well-being of the organization it serves.

Generally, *audit objectives* of both audit groups should be done to:

1. Ensure that an adequate system of controls is implemented and used
2. Determine whether resources are being used in a cost-effective manner
3. Check to see that assets are properly safeguarded and not used improperly
4. Review integrity, reliability, and efficiency of the information system and the financial reports it produces

The auditor's professional judgement dictates how appropriate audit techniques are applied in different audit situations. There are no precise rules to follow as to when and where and in what combination to apply the various techniques. If there were, auditing would not be a professional endeavor but a clerical process.

Burch and Sardinas ([2], pp. 276–283) have described a detailed 10-step procedure to conduct an audit. Figure 11-14 shows a system flowchart using these 10 steps. At the end of the audit the auditor expresses an opinion about the way the business is done in the organization. Four possible opinions are

a. Unqualified
b. Qualified
c. Adverse
d. Disclaimer

An *unqualified opinion* indicates that the audit work was adequate in scope and level of testing, that the system of controls was in place and operating as intended, and that the financial statements fairly present the financial position and results of operation in conformity with generally accepted accounting principles applied on a basis consistent with that of the preceding audit period. Under these conditions, the auditor makes no exceptions and inserts no qualifications in the report.

If the auditor believes that the financial statements do not represent an accurate picture of the company's financial position and operating results or that generally accepted accounting principles have not been applied, then he or she expresses a *qualified opinion*. If, in addition, the company places restrictions on the scope of the auditor's examination, then the auditor forms an *adverse opinion*. Finally, if the restrictions, imposed by the company on the auditor's functions materially limit performance such that he or she cannot comply with the generally accepted accounting principles and cannot apply the computer audit techniques, then the auditor prepares a *disclaimer of opinion*.

The system backup plan discussed in Section 11.17 has a direct bearing on the audit trail and the auditor's functions. An effective backup plan maintains adequate duplicate copies of data and program files. As a result, it becomes simple to trace a transaction from the output to the input stage.

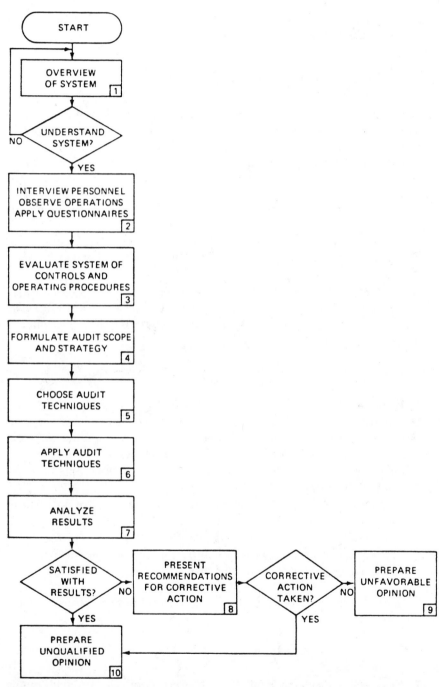

FIGURE 11-14 Flowchart of the audit system process. (Reproduced with permission from J.G. Burch, Jr., and J.L. Sardinas, Jr., *Computer Control and Audit,* Wiley, New York, 1978.)

An auditor can check the reliability and integrity of the data when the system has a good backup plan.

11.19 SYSTEM RECOVERY PLAN

A system recovery plan is a set of procedures to be followed in order to restore a system after a crash (e.g., hardware failure), natural disaster, power outage, human disaster such as riot or vandalism, and so on. The system backup plan may be regarded as a part of the system recovery plan, because in order to restore the system after any type of disaster it is necessary to have duplicate copies of all the data and program files. Thus, the system recovery plan is fairly comprehensive in scope. It belongs to the computer operations department.

As a prerequisite for preparing a system recovery plan, the system team and the computer operations staff should jointly do the following:

a. Review the work being done by the system.
b. Determine which jobs within the system should have priority. In case of a disaster, the operations staff should then restore the higher priority jobs first.

The following steps should be included in a recovery plan:

1. Explore the possibility of manual processing as a stop-gap measure.
2. Use other computer installations with similar hardware and software to run the essential jobs on a temporary basis; i.e., until the complete system is restored. For example, in a distributed processing environment a branch office may be used to run the essential jobs until the head office is back in business.
3. Prepare an agreement with another company in a nearby location to use its facility on a temporary basis.
4. Test the recovery plan on a periodic basis, similar to having fire drills. This ensures that the company will be able to continue business in a nearly normal manner even after disaster occurs.

Various security measures can be taken to avoid possible disasters such as the following:

1. Environmental disaster such as fire, lightning, flood, explosions, earthquakes
2. Electrical or mechanical failure such as power outage, air-conditioning failure, faulty tape drive
3. Theft, i.e., actual stealing of supplies and equipment

4. Fraud in the form of data manipulation or software modification
5. Embezzlement; i.e., theft or fraudulent appropriation of money or other substances entrusted to one's care

Some of these situations can be avoided by taking precautions such as

☐ Environmental protection of an engineering nature such as fire wall, automatic smoke detection system, fire prevention system, automatic sprinklers
☐ Appropriate security measures such as adequate lighting, uniformed guards, closed-circuit TV, alarm system, use of ID badges by employees
☐ Appropriate maintenance of hardware to avoid system crash

11.20   SYSTEM DOCUMENTATION MANUALS

System documentation includes all the descriptions, programs, graphics, and instructions pertaining to the design, implementation, and operation of a system. There are five types of documentation, each addressing a primary audience and described in a corresponding manual. Documentation serves several purposes: During the design phases it describes the evolving product developed by the system team; after implementation , it helps the operations staff run the system, the users use the system, and the programmers modify the system.

The five types of documentation manuals required by a system are

a. System design documentation
b. Software documentation
c. Operations documentation
d. User reference documentation
e. User manual

11.20.1   System Design Documentation

System design documentation consists of the following chapters:

1. Formulation of the problem
2. Feasibility issues
3. Overview of the system, its subsystems and their interfaces
4. Output reports and screen displays
5. Input and source documents
6. Data dictionary

The purpose of the System Design Documentation is to provide technical details of the new system at a top level. Its contents can be derived from the end products of the analysis phase and the preliminary system design phase. Since we are now describing a physical system, the system flowchart is the appropriate graphic tool for enumerating the subsystems and their interfaces. Include detailed specifications of *all* the output reports and *all* the input documents.

11.20.2 Software Documentation

Software documentation consists of the following chapters:

1. Hardware specification: description of the equipment, peripheral devices, communication hardware.
2. System software specification: description of the operating system, all other system software such as DBMS, screen formatter, and forms generator, and communication software.
3. Application software specification: description of each application program, which should include
   ☐ Program objective
   ☐ Input
   ☐ Output
   ☐ Processing logic
   ☐ Hierarchical chart or structure chart
   ☐ Test data
   ☐ Runs and results
   ☐ Source code
4. File design or database design: description of each file used in the system, its access method, and its organization. If a database is used, the complete specifications of the database should be provided. Depending on the complexity of the database, a separate Database Design Documentation may also be prepared.

The overall goal of the software documentation is to help maintenance programmers understand each program and modify the programs.

11.20.3 Operations Documentation

Operations documentation consists of the following chapters:

1. Multiple system flowcharts with supporting narratives: description of each program, its purpose, and the execution sequence of the programs

2. Operation procedure for each program: description of the input required and its formats, files required, processing narrative, output produced, error conditions and operator responses, setup requirements, distribution of output, normal cutoff date for input, and normal run cycle

The overall goal of the operations documentation is to help the operations staff run programs in their proper order and distribute the output. The documentation also tells operators how to handle any error conditions that may appear during the program execution. Whereas the contents of the system design and the software documentations are normally available from the end products of the analysis and the design phases, such is not the case with the operations documentation.

11.20.4 User Reference Documentation

User reference documentation consists of the following chapters:

1. Objective of the documentation: outline of the purpose and use of the document; e.g., user should refer to this document to resolve a problem before contacting the computer staff. A high-quality user reference documentation can reduce the user frustration significantly when the system does not work according to expectation.
2. Input documents: complete list and samples of all the input documents used by the system.
3. Output reports and screen displays: complete list and samples of all the reports, with their frequencies and distributions, and examples of screen displays.
4. Processing logic and error conditions: outline of the processing logic for each output report, error conditions arising out of edit checks at the time of data entry from input documents, and brief description of how the error conditions can be rectified.
5. System maintenance staff: list of computer department staff members who should be contacted in case of any problem that the user cannot rectify.

11.20.5 User Manual

The User Manual consists of the following chapters:

1. Functional capabilities of the system: detailed description, with examples of each functional capability of the system
2. Login procedure and system access: description of the login process and user authorization to access the system, menus, and user prompts for a menu-driven system

3. Generation of output: description of the procedure, with system commands to generate output reports

11.21 IMPORTANCE OF USER MANUALS AND USER REFERENCE DOCUMENTATION

When a new system is implemented, it entails some change in the existing operations. It is human nature to resist changes, even if the changes are for the better. Accordingly, it is quite natural that initially users may offer some degree of resistance or skepticism toward a new system. By involving the user in the system development process, much of this resistance can be removed. Ultimately, a satisfied user community is the greatest supporter of a new system. The user manual and the user reference documentation can win the support of the users, if properly written. Hence these two documents play a great role in the overall "selling" capacity of the new system. Walter Murphy wrote in an article, "User Guide Tells All About System," in *MIS Week* the following:

> A user guide is actually the most revealing part of a system. What a user guide reveals about a system is more important than what the brochure says a system does. The user guide should help you understand what a system does.

Besides the general information about the system given in the introductory sections, there are three basic types of information contained in a user manual:

a. Functional Information, which involves general system operations such as the login procedure, data retrieval from the system, addition/deletion/change of a record

b. Operational Information, which involves information pertaining to the execution of specific functions such as how to respond to an error message, which function keys to press for what actions, which drive a diskette should be loaded into

c. Reference Information, which involves information that is not crucial to run the system but that is occasionally necessary; examples are field descriptions and file format, backup and recovery methods

In general, the functional and operational information appears in the user manual, and the reference information appears in the user reference documentation.

A long-standing criticism against user manuals and user reference documentations is that these are written in an obscure style using computer jargon. As a result, users have a hard time understanding the material. In additon, sometimes the index appears just as a list of acronyms, and users find it very difficult to locate the section(s) addressing their problems.

Any system offers two views: an *inside view* and an *outside view*. The system designer is familiar with the inside view that explains how the system is internally constructed and how it performs the designated functions. Users, on the other hand, are concerned with the outside view, which describes the system as a means of accomplishing certain tasks related to their jobs. A typical user would not ask, What function can the system perform? but instead would ask, How do I accomplish such and such tasks? As a result, if a system designer writes the user manual, which is often the case, then the manual may unknowningly be geared toward the inside view of the system. This will make the manual of little use to users.

In order to remedy this situation, many organizations are now employing technical writers to write the user manuals and user reference documentations. Technical writers can play the role of a user and deal with the outside view of the system. They can also avoid computer jargon as much as possible.

For any large project that I have managed, I always made the technical writer a part of my system design team. Consequently, he or she became familiar with the system development process. At the time of writing the user documentations, the technical writer always put him- or herself in the role of a user and explained how the system performed certain user-designated tasks. The writer always reviewed the work with appropriate system team members to check the accuracy of technical contents. Finally, he or she played the role of a user by trying to use the system on the basis of the user manual alone. This resulted in good quality user manuals and user reference documentations.

11.22 SYSTEM DEVELOPMENT LIBRARY

For a large system development project in which the system team consists of 10 or more members and the project continues for a year or more, it is necessary to keep track of the development process and the volumes of documents generated. Such documents consist of the following:

- ☐ Internal memos and technical reports
- ☐ Graphics such as data flow diagrams, structure charts, HIPO charts, system and program flowcharts
- ☐ Sample input documents, output reports, and screen designs
- ☐ Multiple versions of all the documentations (see Section 11.20)
- ☐ Source codes of all the application software
- ☐ System manuals

and perhaps other documents. Some of these are kept in human readable document form and some in electronic media. Normally, internal memos, technical reports, graphics, and manuals belong to the first category, and the application software and the input data files belong to the second category. When word processors are used, all the memos and reports can also be stored on the electronic media such as diskettes.

A *system librarian* is normally appointed as the custodian of all these documents. He or she starts and maintains the system library, which consists of two parts:

a. Library maintained manually, i.e., human readable form

b. Library maintained in automated form

For (a), the system librarian maintains a list of all the documents generated during the development process. The list is kept in the chronological order of the documents, identifiying for each document its title, author(s), version number, date of publications, and ID number. In addition, an alphabetical index is maintained for quick reference of the documents by titles, and a numerical index is maintained for reference by ID numbers. The system librarian must update the list with the publication of each new document and each new version of an old document.

For (b), the system librarian coordinates the activities of the programmers and the word processing operators. Each programmer is responsible for maintaining his or her codes and notifies the system librarian of the file IDs of these codes. Likewise, the word processing operators generate the reports, manuals, and so on, and notifies the system librarian of the diskette IDs where these documents are stored. Under proper coordination efforts, the system librarian can trace and maintain any document.

11.23   TEXT REVISION CONTROL SOFTWARE

A vareity of software is available that keeps track of revisions of a text. Three such packages are

a. Source Code Control System (SCCS)

b. Software Development Control System (SDCS)

c. Revision Control System (RCS)

These software packages are useful for texts that are revised frequently, for example, programs and documentation. The basis of change between two consecutive versions is tracked by *deltas*, i.e., differences between the two versions. The *grain of change* in any delta is the line. Therefore, if at least one character is changed on a line, then that line forms a part of the delta; i.e., the whole line is considered as changed.

SCCS and SDCS both use merged deltas, which work as follows ([7], p. 62):

Suppose we store the initial revision unchanged and compute the delta for second revision with *diff*. Assume the delta indicates that a single block of lines was changed. Merging the delta into initial revision involves marking the original block of lines as excluded from revision 2 and higher, inserting the block of replacement lines (which may be longer, shorter or empty) right after the first block, and marking the second block as included in revision 2 and higher. Merging additional deltas works analogously, except that excluded and included blocks may overlap. To regenerate a revision, a special program scans through the revision file and extracts all those lines that are marked for inclusion in the desired revision.

Merged deltas have the property that the time for regeneration is the same for all revisions. The whole revision file must be scanned to collect the desired lines. If all revisions are of approximately the same length, the time of copying the desired lines into the output file is also the same for all revisions. Thus, regeneration time is a function of the number of revisions stored and the average length of each revision. However, there is a high cost involved in merging a new delta. First, the old revision must be regenerated to let *diff* compute the delta. Next, the delta is edited into the revision file. This operation is complicated, because it must consider overlapping changes and branches.

RCS uses separate deltas to perform the same functions. It works as follows ([7], pp. 62–63):

Suppose we store the initial revision unchanged. For the second revision, *diff* produces an edit script that will generate the second revision from the first. This script is simply appended to the revision file.

An edit script consists of a sequence of editor commands to delete and insert lines. A deletion command specifies the range of lines to delete, whereas an insertion command specifies after which line to start adding lines, followed by the actual lines to insert. On regeneration, the initial revision is extracted into a temporary file, a simple stream editor is invoked, and the edit script is fed into the editor. This operation regenerates the second revision. Later revisions are stored and regenerated analogously.

The above method applies deltas in a forward direction. The initial revision is stored intact and can be extracted quickly, but all other revisions require the editing overhead. Since the initial revision is accessed much less frequently than the newest one, the deltas should actually be applied in the reverse direction. In such an arrangement, the newest revision is stored intact, and deltas are used to regenerate older revisions. RCS uses this idea. Reverse deltas are not harder to implement than forward deltas, since *diff* generates a reverse delta if the order of its arguments is reversed.

The advantage of separate, reverse deltas is that the revision accessed most often can be extracted quickly—all that is needed is a copy of a portion of the revision file. Regeneration time for the newest revision is merely a function of its length and not of the number of revisions present.

The disadvantage of reverse, separate deltas is that the regeneration of old revisions may take longer than with merged deltas. The problem is that the application of n deltas requires n passes over the text.

The basic distinction among SCCS, SDCS, and RCS is that the first two use forward, merged deltas, whereas the last one uses reverse, separate deltas. Tichy ([7], p. 66) reports that on the basis of an experiment that consisted of timing various operations on a set of benchmark files, he found that an implementation using reverse, separate deltas (e.g., RCS) outperformed one with forward, merged deltas (e.g., SCCS).

11.24 MANAGEMENT OF SYSTEM DEVELOPMENT PROJECT

The analysis, design, and implementation of a system of moderate to high level of complexity is a difficult job. It requires a broad range of talents. Sophisticated techniques of system development and a clear understanding of the detailed user requirements are essential. Consequently, it is necessary to approach the problem in a team environment. The basic elements involved in managing a system team are the task and the team. As the complexity of the project grows, the manager's job becomes the management of the team in such a manner that the team can perform the task properly and complete it on time.

During the analysis phase, the system team remains fairly small and the task is somewhat unstructured. As the work progresses into the preliminary design stage, the size of the team starts to grow, reaching its maximum level during the detailed design and implementation phases. Under the structured programming approach, a single program is decomposed into multiple modules, enabling several programmers to work in parallel on separate modules. It is the responsibility of the team manager to have a global view of the complete system, to break it into several small and manageable tasks, to assign team members to appropriate tasks matching their skills, and to monitor the progress of these tasks so they are integrated into a complete and unified system. This requires a wide variety of talents.

Ideally, a team manager must have both technical and managerial skills, which include excellent interpersonal communications capability. Hugh Carroll commented as follows in his article, "Delegating Responsibility," published in *Electronic Engineering Manager*, February 1985:

He must also be willing to share his authority in order to secure the active cooperation of the people working under him. They must be made to feel that their best efforts are required and that those efforts make a difference. People can only be inventive on their own authority; if the manager's grip is too tight, he can stifle a project's creative flow. At the same time, he must be secure in his overall control of the project, realizing that the full participation of the team will result in a superior product.

The manager must also maintain the highest level of confidence in the managers he supervises directly. He must be aware that they are capable, responsible and are, in fact, candidates for his own job. He has to feel that he can comfortably give away as much of his job as possible in order to concentrate on the progress of the overall task and begin planning for the next project. Delegating authority, however, can be dangerous and must be done in controlled segments.

A system development team normally consists of the following types of people: analysts, designers, programmers, system librarian, technical writer, and secretary. The analysts and the designers are needed during the early part of the work, whereas the programmers are needed during the later part. The system librarian provides support throughout the project (see Section 11.22), as does the secretary. The technical writer is primarily useful during the implementation phase when lots of documentations are produced. The project manager supervises members and is responsible for planning the entire project, scheduling individual tasks within the project, and monitoring the progress throughout the duration of the project.

11.25   GRAPHIC TECHNIQUES FOR PROJECT SCHEDULING

Several graphic techniques are available for scheduling and monitoring a project. Three widely used methods are

  a. PERT (Program Evaluation Review Technique)
  b. MOST (Management Operation System Technique)
  c. LOB (Line of Balance)

CPM, or Critical Path Method, is also used but is really a variation of PERT.
  It is possible to combine some of the aspects of PERT with those of MOST to design an informative network. Such a system can be augmented by using the graphic techniques of LOB, which help top management visualize the current status of a project. Such a combined package can be thought of as a *Generalized PERT System* (see [5]).
  In order to prepare a PERT network for a project, the following steps are needed:

  1. Analyze the project thoroughly and determine the different subprojects or tasks to be accomplished in order to reach the final goal.
  2. Determine the proper sequential order of the various tasks; for example, decide which tasks are to be done first, which tasks next, and so forth.

Most of the time some tasks can proceed simultaneously but some cannot start until certain other tasks are finished. These considerations must be kept in mind in determining the sequential order mentioned in the second step.

In PERT language, each task is called an *activity* or a *job* and the beginning and ending of an activity are called *events*. Simultaneous activities are called *parallel activities*. When several activities have the same beginning event, this event is called a *burst point*; similarly when several activities have the same ending event, this event is called a *node*.

Each event is represented by a circle with a number inside it. The number serves as a convenient reference point for the event. Each activity is represented by a two-line arrow. At the beginning of each activity there is a flag; five figures are written on top of the flag, indicating, respectively, the number of work hours involved in the activity, a cost estimate, and three time estimates of how long the activity will take. In order to keep track of progress, spot checks are made at regular intervals. A vertical line is inserted in the network to indicate when and at what event number the check is made. The date of the spot check is written on the line. If the activity is finished when the spot check is made, the blank space inside the activity arrow is darkened. This shows immediately whether the project is behind, on, or ahead of schedule.

At the top of the each flag you can indicate other relevant data pertaining to the activity. Using three time estimates you can also predict the probability of completing the project on time. The network also can indicate the *critical activities* in the entire project. The critical activities are those that would cause a delay in the completion of the project if any of them were late. There are also *noncritical* activities, which have some *slack time* available; you can delay these activities until their slack time is used up without delaying the project as a whole. The concepts of *early start, early finish, late start,* and *late finish* times are also available in this network.

Along with a generalized PERT network, a LOB graph is designed to represent the progress of the project. This is done by determining what percentage of the whole project each activity entails. The activities are then plotted on a graph showing the time frame. A solid-line graph is used to indicate the *anticipated delivery schedule* and a broken-line graph to indicate the *actual delivery schedule*. A deviation between the two graphs indicates whether the project is behind or ahead of schedule.

Let us now illustrate the technique with an example of a small project consisting of six activities labeled (a) through (f). Figure 11-15 gives the generalized PERT network for the project. Each of the arrows, (a) through (f), represents one activity, and each of the circles, (1) through (8), is an event. Activities (c) and (d) are parallel since they occur simultaneously. Event (3) is a burst point since activities (c) and (d) originate from it. Event (6) is a node since two activities end there. The critical path consists of the events (1)-(2)-(3)-(4)-(6)-(7)-(8), and the total project takes 33–46 working days. The total cost of the project is $1,000, and it requires 782 work hours.

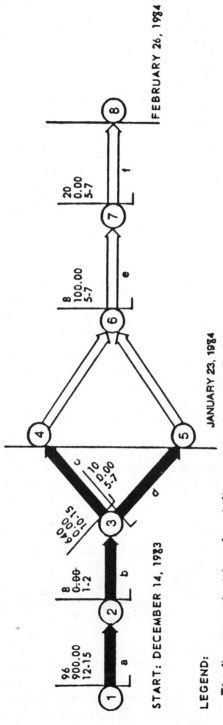

FIGURE 11-15 Generalized PERT network.

START: DECEMBER 14, 1983   JANUARY 23, 1984   FEBRUARY 26, 1984

LEGEND:

The figures against top of each flag represent respectively:

1. The number of manhours involved in the activity.
2. The cost estimate.
3. The most likely time and the most pessimistic time, measured in days.

A spot check is made on January 23, 1984: everything is on schedule. Consequently, the arrows (a) through (d) are darkened.

This is a small project; for large projects with 100 or more activities, you can use a computer to determine the critical path and slack times, similar to using a regular PERT network.

Figure 11-16 is the LOB graph of the project just discussed. The following percentages are assigned to activities (a) through (f), depending on the time needed to perform each of them:

a. 40 percent

b. 15 percent

c. 15 percent

d. 10 percent

e. 10 percent

f. 10 percent

A spot check on January 14, 1984, indicates that the project is running nearly 1 day and 4 percent behind schedule. (The length of the horizontal gap between the end of the actual delivery schedule and the vertical line representing the day of check measures the number of days—or weeks, months, and so forth—that a project is behind or ahead of schedule; similarly, the length of the vertical gap between the end of the actual delivery scheduling curve and the anticipated delivery schedule curve measures what percentage of the total project is running behind or ahead of schedule.) Consequently, the work is accelerated, and by January 23, 1984, the project is back on schedule.

The generalized PERT system is a package consisting of a list of all the activities in the project with percentages assigned to each, a generalized PERT network, and a LOB graph. The advantages of a generalized PERT system include the following:

a. Capability to make direct spot checks at regular intervals of time.

b. Capability to indicate visually, both on the network and on the LOB graph, whether the project is running behind, on, or ahead of schedule.

c. Built-in cost figures to indicate how much the total project will cost. These figures can be revised at the time of each spot check to determine whether the project is running below, above, or in accordance with the budget. If the project is costing more than was originally thought, new estimates can be prepared to indicate the projected cost overrun computed on the basis of the current cost overrun.

d. Capability to indicate what percent of the total and how much actual time the project is behind or ahead of schedule when spot checks are made. The time difference and the difference in the percentage of the project completed can be shown on a graph.

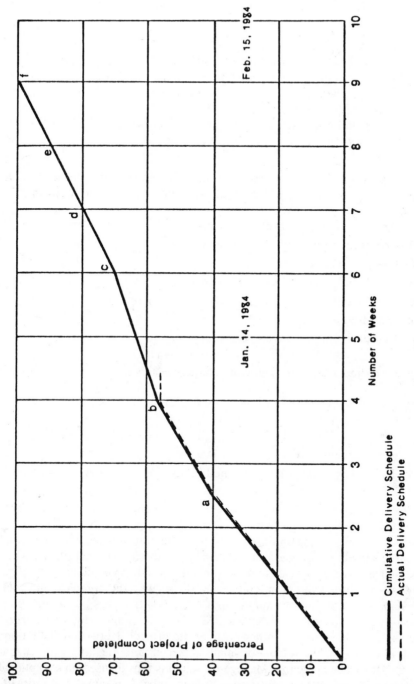

FIGURE 11-16 LOB graph.

e. Capability to indicate how many work hours are needed for each activity.

f. The blank space inside the two-line arrow for an activity can be colored in and a color code developed to indicate which division or department has the responsibility to finish that activity. A color-coded PERT network can be a great visual aid since the responsibilities of different departments are clearly shown. When a color-coded network is used, the space inside each arrow can be covered with black tape when the activity is finished.

The generalized PERT network and the LOB graph can be used for planning and controlling a large variety of projects, including the following:

a. Any research and development project

b. Construction of a building or highway

c. Opening of a new facility

d. Installation and debugging of a computer system

e. Manufacture and assembly of large equipment

f. Design of a training program

g. End-of-the-month closing of accounting records

All such projects have the following characteristics in common:

1. The projects consist of a well-defined collection of activities, which, when completed, mark the end of the project.

2. The activities can be started and stopped independently of each other, with a given sequence.

3. The activities must be performed in a sequential order.

The generalized PERT system is useful at several stages of project management—from the early planning stages, when various alternative programs or procedures are being considered, to the scheduling phase, when time and resource schedules are laid out, and finally to the operational phase, when it is used as a control device to measure actual versus planned progress. The network graph displays in a simple and direct way the complex interrelationships of activities that make up a project. Managers of various subdivisions of the projects may quickly see from the graph how their portion affects, and is affected by, other parts of the project. Network calculations focus attention on the relatively small subset of activities in a project that are critical to its completion. Managerial action is thus focused on exceptional problems—a feature that contributes to more reliable planning and more effective control.

11.26 COMPLIANCE WITH DEADLINES

When a project falls behind schedule, top management normally tries to assign more people to get the project back on schedule. However, this technique does not always work. In fact, depending on the status of the project, sometimes adding extra people may delay the completion of the project. Each project has a saturation point of its own. Once it reaches that point or passes it, increasing the size of the team does not speed up its completion.

The growth of a project is often categorized as *horizontal* or *vertical*. Suppose that a system consists of five subsystems. If the system development team builds only three subsystems, an additional group of people may be employed to build the fourth subsystem and to interface with the other three subsystems being developed. This is considered a horizontal growth. Employing additional team members can quicken the completion time in case of a horizontal growth. On the other hand, if the project involves a conversion of a prototype into an operational system, then comprehensive knowledge of the prototype is essential. For example, the prototype may include implementation of only three edit criteria, whereas the full system needs a total of seven edit criteria. Such a growth is labeled vertical growth. A team member cannot work in a vertical growth environment without knowing what has already happened in the prototype. Consequently, in such cases, adding extra people does not normally help to speed up the completion of the project. Figure 11-17 depicts the approximate relationship between the addition of personnel to a project and the completion time of the same.

Frederick Brooks has commented as follows regarding the above topic ([1], pp. 16–18):

> The man-month as a unit for measuring the size of a job is a dangerous and deceptive myth. . . . Men and months are interchangeable only when a task can be partitioned among many workers with no communication among them. . . . The added burden of communication is made up of two parts, training and intercommunication. Each worker must be trained in the technology, the goals of the effort, the overall strategy, and the plan of work. This training cannot be partitioned, so this part of the added effort varies linearly with the number of workers.

Brooks has even offered the following Brooks' Law ([1], p. 25):

> Adding manpower to a late software project makes it later.

11.27 INTERPERSONAL CONFLICTS IN TEAM ENVIRONMENT

When a group of people work together with a common goal such as a system development project, occasional interpersonal conflicts show up. The project

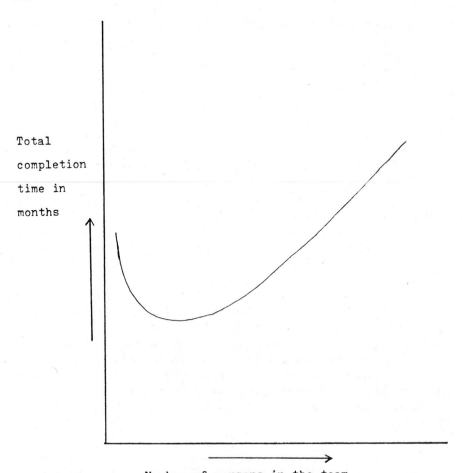

FIGURE 11-17   Staffing level versus project completion time.

manager tries to keep such outbursts under control. He or she arranges weekly meetings for discussing problems, reexamining goals and probably rescheduling tasks, and attempting to make the maximum use of available time and skills of the team members. But at times such meetings may appear as the occasion for venting out frustrations. The project manager must be skillful enough to control this situation.

Wetherbe ([8], pp. 235–236) has offered some general guidelines to minimize disruptions caused by interpersonal conflicts or dysfunctional behavior:

a. A malfunctioning system may cause frustrations among the team members. So adequate communication should exist among the team members to pinpoint deficiencies in the development approach.

b. An extra tight schedule can lead to enormous stress among team members. The project leader should schedule the project carefully and allow enough slack times in the PERT network of the project. The project leader should not yield to pressures exerted by top management to enforce an unrealistic schedule. As Brooks ([1], p, 21) puts it:

> Individual managers will need to stiffen their backbones and defend their estimates with the assurance that their poor hunches are better than wish-derived estimates.

Let us conclude this topic with the following upbeat comment by Hugh Carroll published in his paper, ''Delegating Responsibility'':

> A team with a clear sense of purpose—that shares responsibility and credit, a team where the manager sets an example and instills a feeling of success—is a team that will perform at a consistently high level and will show the ability to surmount the unexpected crises with deceptive ease.

11.28 SYSTEM MAINTENANCE

Upon completion of system conversion (see Section 11.16), the system changes from the development stage to the operational stage. However, most information systems require at least some modifications after development. This need arises from a failure to anticipate all requirements during systems design and/or from changing organizational requirements. The second of these, changing requirements, continues to have an impact on most information systems as long as they are in operation. Consequently, periodic systems maintenance is a requirement of most information systems. Systems maintenance involves adding new data elements, modifying reports, adding new reports, changing calculations (e.g., payroll tax tables), and the like.

There are two types of system maintenance:

a. Preventive maintenance
b. Rescue maintenance

Preventive maintenance refers to prescheduled maintenance on a regular, say monthly, basis. This is anticipated and is planned for in advance. Rescue maintenance refers to on-demand maintenance. It pertains to previously undetected malfunctions that are not anticipated but that require immediate attention. A system that is properly developed and tested should have few occasions of rescue maintenance.

The primary responsibility of the system development team is over when the system becomes operational. However, it is advisable and often necessary to establish a continuing relationship between the customer and the development team. In such a situation, a certain fraction of the team of development programmers may be retained on an on-call basis for maintenance works, primarily rescue maintenance. If the customer's organization has a computer operations department, its employees can take care of preventive maintenance.

11.29 SYSTEM EVALUATION

The final step in the system development process is system evaluation. Evaluation provides the feedback needed to access the value of the information provided by the system. It addresses the basic question, Does the new system perform as envisaged? The answer to this question provides guidance about what adjustments are needed to make the information system perform according to its desired goal. The evaluation process concerns the quality of the information generated by the system.

Normally, system evaluation is done through interviews and distribution of questionnaires among the system users. This is a subjective process and cannot be conducted in a straightforward quantitative manner. The interviews with system users regarding the performance of the new system should be done so as to preserve anonymity. The questionnaires should be properly designed to elicit the necessary response and then should be tabulated to arrive at the final opinion of the users.

For any system there are two categories of users: regular users who *have* to use it for their daily functions and occasional users who *want* to use it if the effort is worthwhile. This latter group may be called voluntary users. If a new system has a large group of voluntary users, it may be concluded that the system performs satisfactorily. Also, the regular users can comment on the "user friendliness" of the system. A large group of satisfied users is always the greatest testimonial of a new system.

In order to evaluate a system from an objective operational viewpoint, the following questions may be asked ([8], p. 201):

1. Are all transactions processed on time?
2. Are all values computed accurately?
3. Is the system easy to work with and understand?
4. Is terminal response time within acceptable limits?
5. Are reports processed on time?
6. Is there adequate storage capacity for data?

Responses to these questions determine the quality of system operation.

An information system may remain in an operational and maintenance mode for several years. The system should be evaluated periodically to ensure that it is operating properly and is still proper for the organization. When a system becomes a problem (i.e., no longer satisfies the organization's need) or new opportunities are available (e.g., new technology), an information system may be replaced by a new information system. This leads to the problem definition phase and starts the system life cycle over again.

11.30 END PRODUCT OF SYSTEM IMPLEMENTATION PHASE

The system implementation phase is the last part of the system life cycle. Accordingly, its end product must wrap up the whole process. The following items should be included in this end product:

1. **System flowcharts for the complete system and each subsystem**, which provides a continuation of the detailed design phase. A brief narrative should accompany the diagrams describing the operations of the subsystems

2. **Complete documentation of each program**, consisting of the following:
   a. Input/output table
   b. Structure chart or hierarchy chart
   c. IPO charts, one for each module of the program
   d. Input file(s)
   e. Source code listing

Such a package describes the detailed structure of each program as a hierarchy of its component modules. The source code must match this structure.

3. **Plan for physical site preparation**, which will be used to prepare the installation site before system implementation.

4. **Plan and schedule for user training**, which ensures that the users are adequately trained to use the system. The schedule should be monitored closely for compliance with training deadlines.

5. **Plan for system conversion**, which should specify the type of conversion—discrete or parallel. For parallel conversion, it should specify if it is to be implemented as a partial conversion or a complete conversion. Monitoring the plan is of utmost importance.

6. **System backup plan**, which describes how the new system will be backed up so as not to lose data. A direct consequence of this plan is the capability to establish an audit trail.

7. **System recovery plan**, which describes how the system can be restored to its normal operating state in case of a disaster—natural or accidental. The plan should be tried on a regular basis to check its feasibility.

8. **Documentation and user manuals,** which consists of a series of manuals that describe the design and operation of the system and also tell users how to run the system for getting information from it (see Section 11.20). The documentation manuals form a vital part of the saleability of the system.

11.31 SUMMARY

This chapter discussed the last phase of the system life cycle—system implementation, maintenance, and evaluation. The topics covered were divided into three categories:

a. Program level detailed design
b. System implementation
c. System maintenance and evaluation

The program level design part describes the tools needed for the design and documentation of structured programs. The input/output table shows, in a list format, the input and output of each program mentioned in the system flowcharts. The exact hierarchical top-down structure of each program can be described by using structure charts or HIPO charts. The structure of each module in a program is explained by an IPO chart or program flowchart. The former uses structured English, and the latter uses graphic technique to describe the logic and flow of control of each module. Pseudocodes are sometimes used to describe detailed contents of each program. However, a well-designed IPO chart can eliminate the necessity of pseudocodes.

The system implementation part includes topics related to the activities for making the system operational; i.e., transforming the "paper" system into an electronic system. The techniques of structured programming are discussed, explaining that a program should use the three constructs—sequence, decision, and repetition—and should avoid GOTOs as far as practicable. Some basic principles of program testing and debugging follow. Before installation of hardware, it is necessary to prepare the installation site. Thus, a brief description of that activity is given. The next broad topic is user training, which describes how training should be conducted, who should be the participants, and what should be the schedule. Sound user training is needed for a smooth operation of the new system.

System conversion is the process of transition from the old system to the new system. This can happen in a direct one-time approach or in a phased gradual approach. The latter is normally preferred. The system backup plan describes how data and program files should be copied and saved for future use, especially to track a transaction. This process ensures the establishment of an audit trail for each transaction. The latter is used by an auditor during a data processing audit. The system recovery plan formulates procedures to restore the system to a running condition after a disaster—natural or human. Some of these procedures relate to security, such as restricting the flow of visitors and installing environmental protection measures.

System documentation consists of a series of five documents that describe in detail the design of the system and the software, the operational procedures for running the system, and the user manuals. These documents should be prepared carefully because they determine how users of the new system will receive it. A satisfied user community is of utmost importance for the post-implementation success of a new system.

System management addresses some major issues pertaining to the management of a system development project. It outlines the role of the project manager and gives a tentative composition of the team. A graphic tool called generalized PERT system was described. It can be used for planning, scheduling, and monitoring a system development project. The system team should include a system librarian who keeps track of all the development details and assists in the preparation of system documentations. The issue of possible interpersonal conflicts among the team members was addressed briefly.

System maintenance and evaluation relate to activities after the system becomes operational. Maintenance includes both preventive and scheduled maintenance and rescue maintenance on demand. A small fraction of the system development team should be retained for rescue system maintenance, especially for program modifications. System evaluation is done after the system has been operational for at least 6 months. It examines the quality, relevance, and usefulness of the information generated by the system. Users' opinion and comments are collected through interviews and questionnaires.

The chapter closed with a list of items that are normally included in the end product of this phase.

11.32 KEY WORDS

The following key words are used in this chapter:

abstraction	cohesion
activity	cohesiveness
actual delivery schedule	complete parallel conversion
adverse opinion	control module
afferent branch	coupling
anticipated delivery schedule	critical activity
audit objective	dated relation
audit trail	debugging
Brooks' Law	decision
burst point	delta
casual user	disclaimer of opinion

discrete conversion
early finish time
early start time
efferent branch
event
generalized PERT system
grain of change
heavy user
hierarchy
HIPO chart
horizontal growth
input/output table
IPO chart
late finish time
late start time
Line of Balance (LOB)
link test
Management Operation System Technique (MOST)
merged delta
module
node
parallel activity
parallel conversion
partial parallel conversion
preventive maintenance
Program Evaluation Review Technique (PERT)
program flowchart
program level design
pseudocode
qualified opinion
repetition
rescue maintenance
Revision Control System (RCS)

scope of control
scope of effect
separate delta
sequence
size
slack time
Software Development Control System (SDCS)
Source Code Control System (SCCS)
string test
structure chart
structured English
structured programming
stub
stub test
system backup
system conversion
system development library
system documentation
system evaluation
system implementation
system level design
system librarian
system maintenance
system recovery
system's inside view
system's outside view
three-generation backup method
transform
unqualified opinion
user manual
user training
vertical growth

REFERENCES

1. F. P. Brooks, Jr., *Mythical Man-Month*, Addision-Wesley, Reading, MA, 1975.
2. J. G. Burch, Jr., and J. L. Sardinas, Jr., *Computer Control and Audit*, Wiley, New York, 1978.

3. W. S. Davis, *Systems Analysis and Design: A Structured Approach*, Addison-Wesley, Reading, MA, 1983.

4. R. W. Jensen and C. C. Tonies, *Software Engineering*, Prentice-Hall, Englewood-Cliffs, NJ, 1979.

5. S. S. Mittra, "PERT, LOB, and MOST: United for More Efficient Project Scheduling," *Supervisory Management*, 30–35 (November 1976).

6. D. Prather, *Discrete Mathematical Structures for Computer Science*, Houghton-Mifflin, Boston, 1976.

7. W. F. Tichy, "Design, Implementation and Evaluation of a Revision Control System," *IEEE Proceedings of Sixth International Conference on Software Engineering*, 58–87 (1982).

8. J. C. Wetherbe, Jr., *System Analysis for Computer-Based Information Systems*, West Publishing, St. Paul, MN, 1979.

REVIEW QUESTIONS

1. Describe briefly the three main components of the system implementation phase.
2. What purpose is served by an input/output table?
3. Explain, with examples, the principal characteristics of a structure chart.
4. Why is it necessary that coupling should be low and cohesion should be high?
5. Describe the functions of the three branches of a structure chart.
6. What is the distinction between a structure chart and a hierarchy chart? Which one do you prefer and why?
7. Given a choice between an IPO chart and a program flowchart to document a program, which one would you select and why?
8. Do you think that pseudocode is necessary? Is a pseudocode syntax specific for a language? That is, does the pseudocode for a FORTRAN program differ from that for a COBOL program? Justify your answer.
9. What are the three basic constructs in a structured program? How does DO WHILE differ from DO UNTIL?
10. How can you avoid using GOTOs in a program?
11. Is a GOTO-less program the same as a structured program?
12. Define the terms stub test and string test.
13. For what types of system development projects do you need a plan for preparation of physical site?
14. Explain why user training is so vital for a system development project.
15. Describe the different types of system conversion. Which one do you prefer and why?
16. Explain the statement, Backup is a form of insurance.
17. What is the purpose of a data processing audit? What are the four possible opinions that an auditor can include in the audit report?
18. How does a system recovery plan differ from a system backup plan?

19. Describe the different types of documentation needed for a system.
20. Why is it so important that user manuals be properly done?
21. Distinguish between the user's view and the designer's view of a system. Which documentation emphasizes which view?
22. Describe the function of a system librarian.
23. Explain the difference between forward merged deltas and reverse separate deltas.
24. Is it always true that adding people to a project speeds up the implementation? What is Brooks' law? Do you agree with its contention?
25. How should a project manager handle interpersonal conflicts among the team members?
26. Describe the two types of system maintenance.
27. Why is it necessary to conduct a system evaluation? How does one conduct it?

FORMULATIONAL PROBLEM

Select a project that you have developed as team member. Assume that you have used the structured methodology throughout. Now answer the following questions:

a. Select a process from a microlevel DFD and find its description in the data dictionary. How does this description relate to the IPO chart(s) of the program(s) generated by the process?
b. Select a functional requirement of the system. Describe how that function is implemented by tracing it through the system from the DFD to the structure chart.

12

System Implementation, Maintenance, and Evaluation: Case Studies

12.1 INTRODUCTION

In this chapter we shall discuss the system implementation phase of the two case studies—Toy World, Inc., and Massachusetts Educational Foundation. This will conclude the system development process for both of them. For each case study, we start with the system flowchart given in Chapter 10 and complete the system implementation phase by following the theoretical framework developed in Chapter 11.

12.2 SYSTEM IMPLEMENTATION FOR TOY WORLD

12.2.1 System Flowchart for Selected Subsystem

Section 10.2.1 described the order processing subsystem built around the process of customer change transactions. Figure 12-1, which is essentially the same as Figure 10-1, shows the seven programs that constitute the subsystem, namely,

1. CRC410   Accept customer change transactions
2. CRC415   Backup program

3.	CRC420	Disperse transactions
4.	CRC430	Update master file
5.	CRC440	Update sales file
6.	CRC450	Update customer file
7.	CRC460	Update inventory file

Figure 12-1 shows the input and the output for each of these programs.

12.2.2   Input/Output Tables

Figure 12-2 shows the input/output table for the seven programs. For each program, the figure shows the input and the output. The output may consist of hardcopy reports or files. Some of the input to a later program are derived from the output of an earlier program. For example, the input to CRC415 is derived from the output of CRC410.

12.2.3   Hierarchy Charts

HPC has decided to use a hierarchy chart and a program flowchart for each program listed in Figure 12-2. Together, these two charts show the complete structure and logic flow of each program. The hierarchy charts follow the standard principle of separating the flow of control into three branches, namely, afferent, transform, and efferent. Figures 12-3 through 12-16 give the seven hierarchy charts and the seven program flowcharts for the seven programs, CRC410 through CRC460. They complete the program level design of the selected subsystem.

12.2.4   System Backup Plan

The system backup plan consists of backing up all data and program files. The company presently has no means of system backup. HPC recommends that all major files, i.e., master file, sales file, customer file, and inventory file, be backed up. This will ensure quick uptime if the system were to crash. We recommend two daily backups during normal times and three backups during peak seasons. We also recommend two separate tapes for each backup, one to be kept on site for immediate recovery (within 1 hour) and the other to be moved immediately to an off-site storage area. This way, in the event of a major site disaster, these tapes will be preserved for that phase of the recovery plan.

Magnetic tape has been selected as the backup medium because it is ideal for this size system. The expense incurred is more cost effective even though it takes longer to accomplish in comparison with using disk as the backup medium.

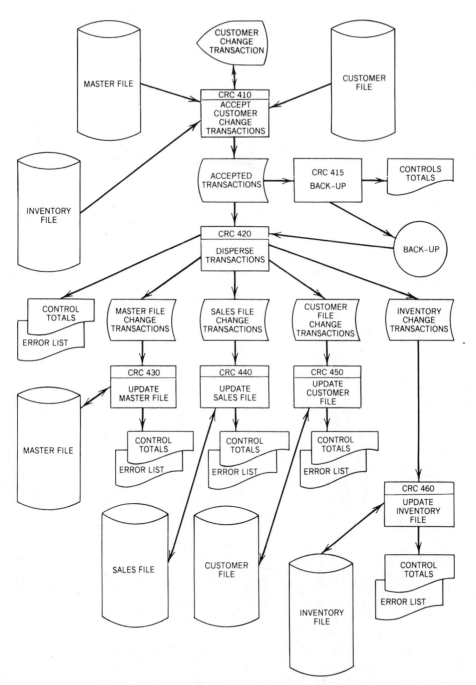

FIGURE 12-1 System flowchart for a selected subsystem.

Input/Output Table

Input	Program	Output
1. On-line entered customer requested change data 2. Master file 3. Customer file 4. Inventory file	CRC410 (Accept Customer Change Transactions Program)	1. Customer requested change transactions
1. Customer requested change transactions from CRC410	CRC415 (Backup Program)	1. Customer requested change transactions backup tape 2. Control totals
1. Customer requested change transactions from CRC410 or CRC415	CRC420 (Disperse Transaction Program)	1. Master file change transactions 2. Sales file change transactions 3. Customer file change transactions 4. Inventory file change transactions 5. Control totals 6. Error list
1. Master file change transactions from CRC420 2. Master file	CRC430 (Update Master File Program)	1. Updated Master File 2. Control totals 3. Error list

Input	Program	Output
1. Sales file change transactions from CRC420 2. Sales file	CRC440 (Update Sales File Program)	1. Updated Sales File 2. Control totals 3. Error list
1. Customer file change transactions from CRC420 2. Customer file	CRC450 (Update Customer File Program)	1. Updated Customer File 2. Control totals 3. Error list
1. Inventory file change transactions from CRC420 2. Inventory file	CRC460 (Update Inventory File Program)	1. Updated Inventory File 2. Control totals 3. Error list

FIGURE 12-2 Input/output table.

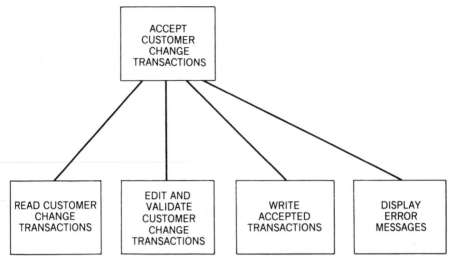

FIGURE 12-3 Hierarchy chart for CRC 410.

12.2.5 System Recovery Plan

In order to minimize loss due to accident, hazard, malfunction, fraud, and/or theft, the following is recommended:

As a minimal requirement for any type of recovery, backup files should be maintained and stored off site for all critical files. In addition, in order to reduce the potential for loss, a preventive maintenance program should be set up and maintained for hardware and software.

To recover loss of equipment in the event of an accident or hazard, an insurance policy should be purchased covering the cost of equipment.

For a short down time involving the operability of the hardware or software, the vendor should be contacted by key person in the computer installation (preferably the computer operations manager). Service will be provided as a part of hardware/software requisition.

For a long down time in which partial or complete loss of computer facilities occurs, the following plan should be implemented:

1. Key personnel should be established to implement the plan.
2. Any and all programs, master files, and documents that can be saved should be saved.
3. A contract should be established and maintained with local time sharing vendor with compatible equipment. This equipment will be used to run the company's critical workload. As such, critical systems should be identified in order to keep operations ongoing.

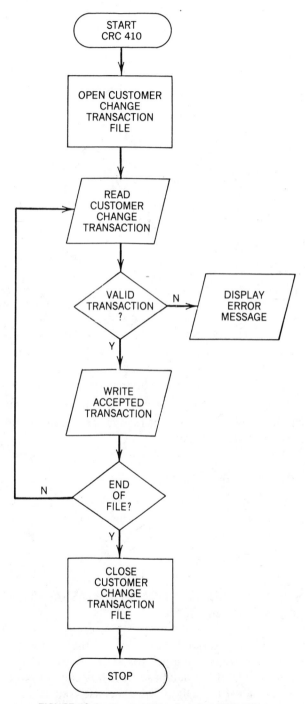

FIGURE 12-4 Program flowchart for CRC 410.

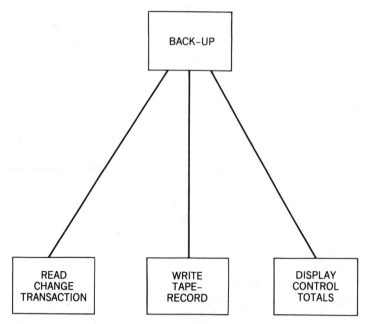

FIGURE 12-5   Hierarchy chart for CRC 415.

4. Noncritical reports should be eliminated if possible.
5. Amount of time needed to run critical workload should be established.
6. Length of recovery and cost of recovery method should be estimated to anticipate potential risks.

12.2.6 Source Code Listing

Figure 12-17 gives the source code in COBOL of the program CRC430, Update Master File. Figure 12-18 is an error list produced by running CRC430. These figures are provided as samples of the source code listings and associated output.

12.3 SYSTEM IMPLEMENTATION FOR MASSACHUSETTS EDUCATIONAL FOUNDATION

In Section 10.3.1 we described the details of the database design for the financial reporting system developed for MEF. In this section, we give the implementation details of the database, which is called the Form Database.

As noted in Section 10.3.1, the Form Database contains 25 relations called MEF-XX, where XX ranges from 01 to 25, one relation called DB-STATUS,

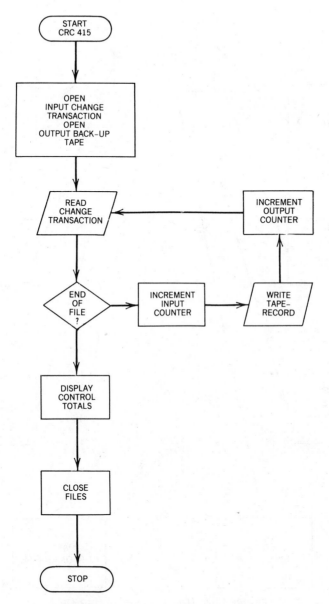

FIGURE 12-6 Program flowchart for CRC 415.

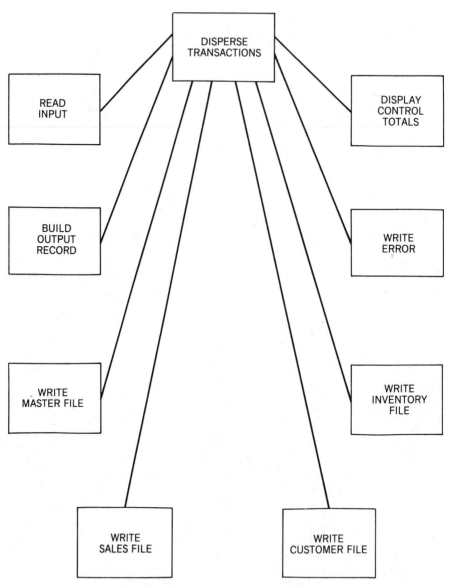

FIGURE 12-7 Hierarchy chart for CRC 420.

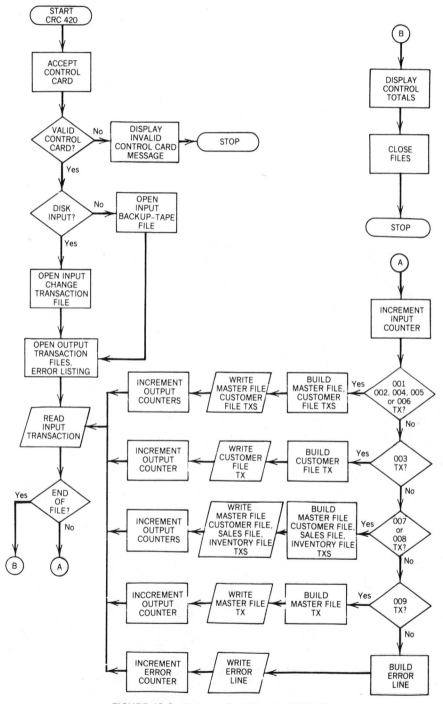

FIGURE 12-8 Program flowchart for CRC 420.

319

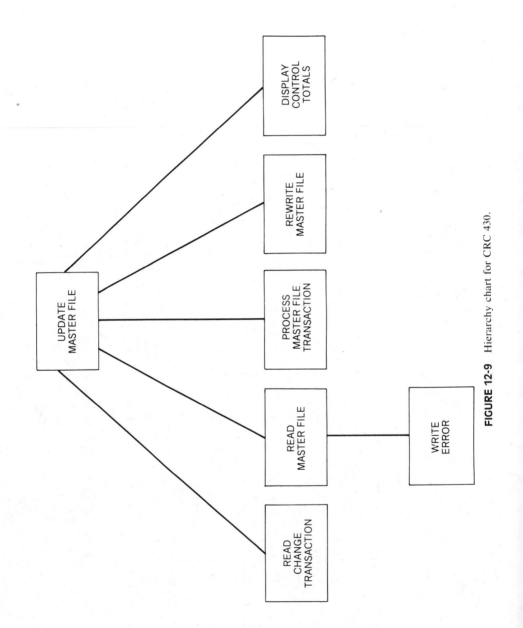

FIGURE 12-9 Hierarchy chart for CRC 430.

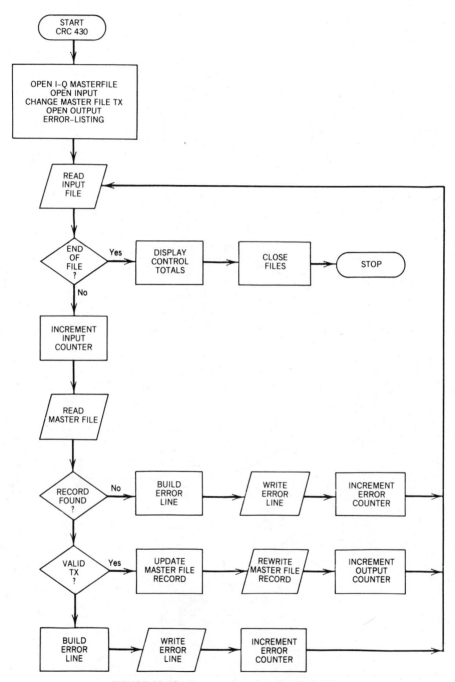

FIGURE 12-10 Program flowchart for CRC 430.

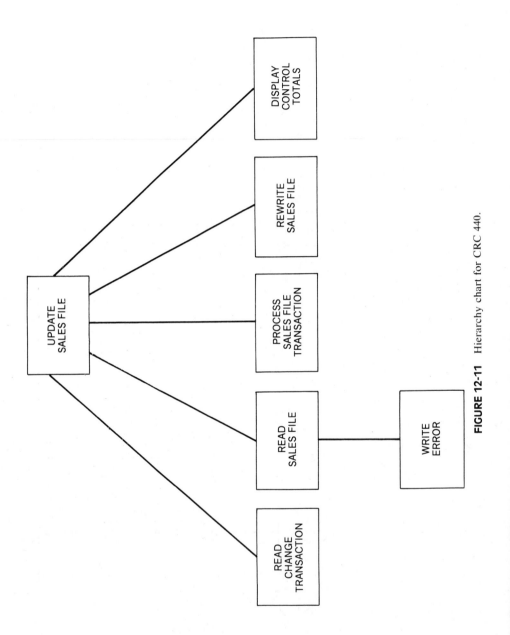

FIGURE 12-11 Hierarchy chart for CRC 440.

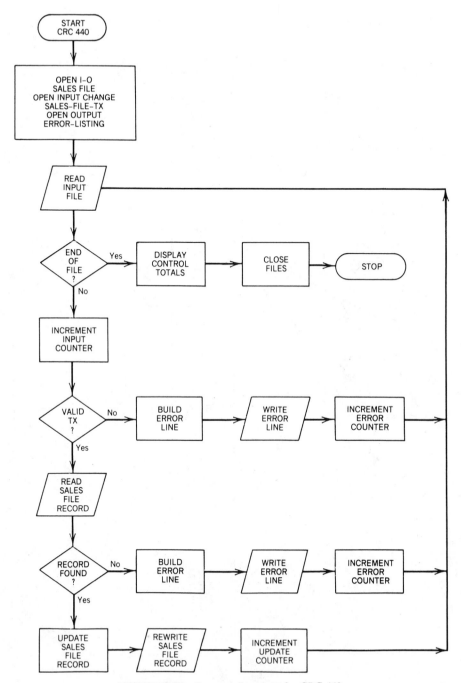

FIGURE 12-12 Program flowchart for CRC 440.

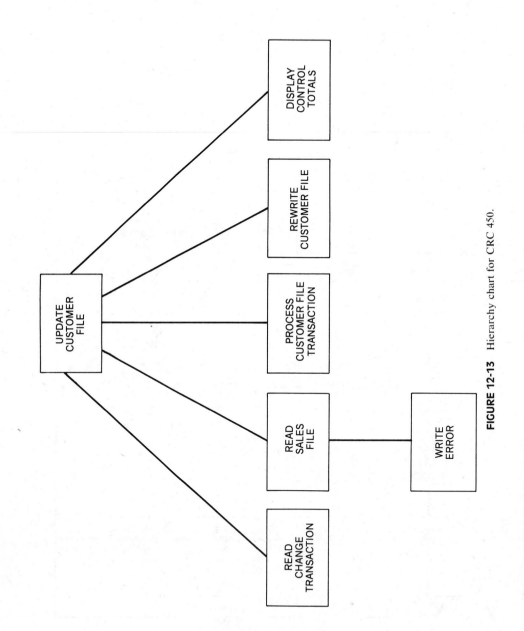

FIGURE 12-13 Hierarchy chart for CRC 450.

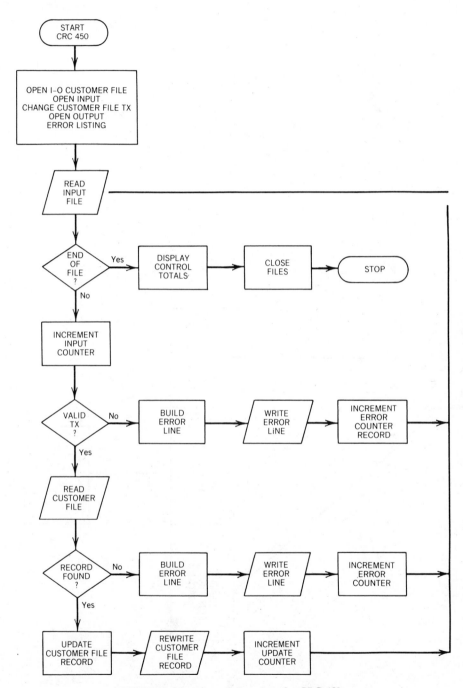

FIGURE 12-14 Program flowchart for CRC 450.

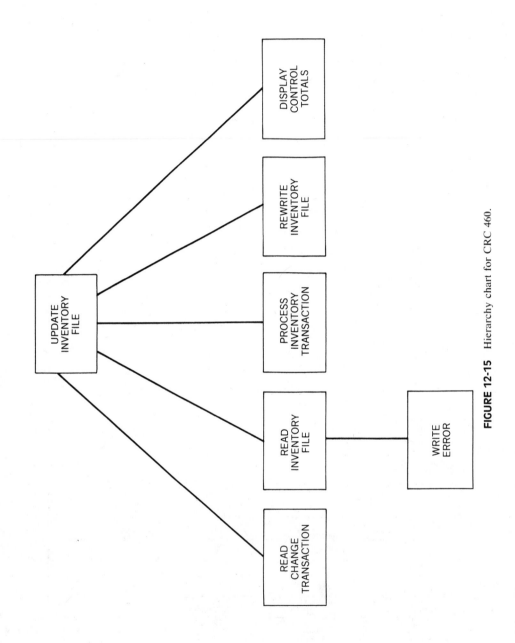

FIGURE 12-15 Hierarchy chart for CRC 460.

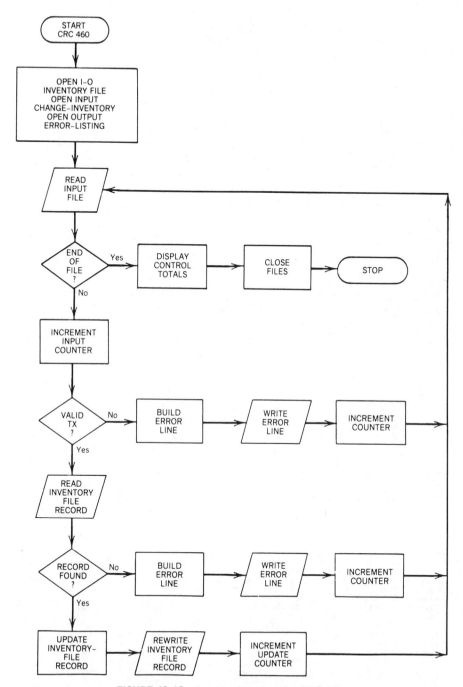

FIGURE 12-16 Program flowchart for CRC 460.

```
CBL    BUF#06230,SEQ,FLAGW,NOSUPMAP,SPACE1,LVL#NO,NOCLIST
CBL    NOSTXIT,APOST,ZWB,NOSYNTAX,NOSXREF,PMAP#0,NOOPT
CBL    NOSTATE,NOCATALR,NOLIB,NOVERB,NOVERBSUM,NOVERBREF
CBL    NOCOUNT,NOTRUNC
CBL    LIB,CLIST,SXREF,STATE,FLOW                                    0000001(

00002    000030 IDENTIFICATION DIVISION.
00003    000040 PROGRAM-ID. CRC430.
00004    000050 AUTHOR. HPC.
00005    000060 DATE-COMPILED. 12/01/83.
00006    000070 DATE-WRITTEN. 11/83.
00007    000080 REMARKS.    CRC430 WILL READ AS INPUT THE CUSTOMER REQUESTED
00008    000090            CHANGE TXS, OUTPUT OF CRC420, USED TO UPDATE THE
00009    000100            MASTERFILE.
00010    000110            AN ERROR LIST WILL BE GENERATED TO DISPLAY INVALID
00011    000120            TXS AND CONTROL TOTALS WILL BE PRODUCED WHEN PROCESSING
00012    000130            IS COMPLETED.

00014    000150 ENVIRONMENT DIVISION.
00015    000160 CONFIGURATION SECTION.
00016    000170 SOURCE-COMPUTER. IBM-4341.
00017    000180 OBJECT-COMPUTER. IBM-4341.
00018    000190 INPUT-OUTPUT SECTION.
00019    000200 FILE-CONTROL.
00020    000210     SELECT MASTERFILE            ASSIGN TO SYS001-MSFILE
00021    000220                                  FILE STATUS IS STATUS-VS
00022    000230                                  ORGANIZATION IS INDEXED
00023    000240                                  ACCESS IS DYNAMIC
00024    000250                                  RECORD KEY IS MF-INVOICE-ORDER-NO.
00025    000260     SELECT CHANGE-MASTERFILE-TX ASSIGN TO SYS002-DA-3350-S.
00026    000270     SELECT ERROR-LISTING         ASSIGN TO SYS003-UR-1403-S.
00027    000280
00028    000290*** SYS000 IS USED IN THE JCL TO REFERENCE THE MASTERFILE


      2        CRC430          14.25.21        12/01/83


00030    000310 DATA DIVISION.
00031    000320 FILE SECTION.
00032    000330 FD  MASTERFILE
00033    000340     LABEL RECORDS ARE STANDARD
00034    000350     RECORD CONTAINS 137 TO 560 CHARACTERS
00035    000360     BLOCK CONTAINS 1374 TO 5604 CHARACTERS
00036    000370     DATA RECORD IS MASTERFILE-RECORD.
00037    000380
00038    000390 01  MASTERFILE-RECORD.
00039    000400     03  MF-INVOICE-ORDER-NO          PIC S9%08□     COMP-3.
00040    000410     03  MF-CUSTOMER-NO               PIC S9%06□     COMP-3.
00041    000420     03  MF-CUSTOMER-NAME             PIC X%16□.
00042    000430     03  MF-CUSTOMER-ADDRESS.
00043    000440         05  MF-CUSTOMER-STREET       PIC X%15□.
00044    000450         05  MF-CUSTOMER-CITY-STATE   PIC X%20□.
00045    000460         05  MF-CUSTOMER-ZIP          PIC S9%05□     COMP-3.
00046    000470     03  MF-DATE-ORDER-RECEIVED       PIC S9%07□     COMP-3.
00047    000480     03  MF-DATE-SHIPMENT-REQUESTED   PIC S9%07□     COMP-3.
00048    000490     03  MF-SALESPERSON-NO.
00049    000500         05  FILLER                   PIC X%04□.
00050    000510         05  MF-DISTRICT              PIC X%01□.
00051    000520             88  MF-DISTRICT-0001                   VALUE @A@.
00052    000530             88  MF-DISTRICT-0002                   VALUE @B@.
00053    000540             88  MF-DISTRICT-0003                   VALUE @C@.
00054    000550             88  MF-DISTRICT-0004                   VALUE @D@.
00055    000560             88  MF-DISTRICT-0005                   VALUE @E@.
00056    000570             88  MF-DISTRICT-0006                   VALUE @F@.
00057    000580             88  MF-DISTRICT-0007                   VALUE @G@.
00058    000590     03  FILLER                       PIC X%08□.
00059    000600     03  MF-TOTAL-COST                PIC S9%05□V99  COMP-3.
00060    000610     03  MF-TOTAL-ENTRIES             PIC S9%02□.
00061    000620     03  MF-ORDER-ENTRIES   OCCURS 10 TIMES
00062    000630                            DEPENDING ON MF-TOTAL-ENTRIES.
00063    000640         05  MF-STYLE-CATEGORY        PIC X%02□.
00064    000650         05  MF-STYLE-NO-ORDERED      PIC S9%05□     COMP-3.
00065    000660         05  MF-BIN-LOCATION          PIC X%07□.
00066    000670         05  MF-QUANTITY-ORDERED      PIC S9%05□     COMP-3.
00067    000680         05  MF-QTY-SHIPPED           PIC S9%05□     COMP-3.
00068    000690         05  MF-TOTAL-PRICE           PIC S9%05□V99  COMP-3.
00069    000700         05  MF-ENTRY-STATUS          PIC X%01□.
00070    000710             88  MF-NOT-PROCESSED                   VALUE @A@.
```

FIGURE 12-17 Source code listing of CRC 430.

328

```
00071   000720              88    MF-ORDER-FILLED                           VALUE @B@.
00072   000730        05    MF-ORDER-STATUS          PIC   X%01¤.
00073   000740              88    MF-A-OK                                   VALUE @A@.
00074   000750              88    MF-FULL-SUBSTITUTION                      VALUE @B@.
00075   000760              88    MF-PARTIAL-SUBSTITUTION                   VALUE @C@.
00076   000770              88    MF-FULL-BACK-ORDER                        VALUE @D@.
00077   000780              88    MF-PARTIAL-BACK-ORDER                     VALUE @E@.
00078   000790              88    MF-FULL-CANCELLATION                      VALUE @F@.
00079   000800              88    MF-PARTIAL-CANCELLATION                   VALUE @G@.
00080   000810        05    MF-SUBSTITUTED-STYLE-NO   PIC   S9%05¤     COMP-3.
00081   000820        05    MF-QTY-SUBSTITUTED        PIC   S9%05¤     COMP-3.
00082   000830        05    MF-SUBS-BIN-LOCATION      PIC   X%07¤.


      3       CRC430              14.25.21        12/01/83


00083   000840        05    MF-QTY-BACK-ORDERED       PIC   S9%05¤     COMP-3.
00084   000850        05    FILLER                    PIC   X%07¤.
00085   000860
00086   000870 FD  CHANGE-MASTERFILE-TX
00087   000680     RECORDING MODE IS V
00088   000890     LABEL RECORDS ARE STANDARD
00089   000900     RECORD CONTAINS 79 TO 239 CHARACTERS
00090   000910     BLOCK CONTAINS 794 TO 2394 CHARACTERS
00091   000920     DATA RECORD IS CHANGE-MF-TX.
00092   000930
00093   000940 01  CHANGE-MF-TX.
00094   000950        03    CMF-COMMON-DATA.
00095   000960        05    CMF-TX-NUMBER            PIC   S9%05¤     COMP-3.
00096   000970        05    CMF-TX-DATE             PIC   S9%06¤     COMP-3.
00097   000980        05    CMF-TX-CODE             PIC   X%03¤.
00098   000990              88    CMF-NAME-TX                              VALUE @001@.
00099   001000              88    CMF-ADDR-TX                              VALUE @002@.
00100   001010              88    CMF-NAME-ADDR-TX                         VALUE @004@.
00101   001020              88    CMF-NAME-PHONE-TX                        VALUE @005@.
00102   001030              88    CMF-ADDR-PHONE-TX                        VALUE @006@.
00103   001040              88    CMF-CANCEL-ORDER-TX                      VALUE @007@.
00104   001050              88    CMF-ALTER-ORDER-TX                       VALUE @008@.
00105   001060              88    CMF-SHIPMENT-DATE-TX                     VALUE @009@.
00106   001070        03    CMF-DATA.
00107   001080        05    CMF-INVOICE-NO          PIC   S9%08¤     COMP-3.
00108   001090        05    CMF-CUSTOMER-NO         PIC   S9%06¤     COMP-3.
00109   001100        05    CMF-NAME                PIC   X%16¤.
00110   001110        05    CMF-ADDRESS.
00111   001120              07    CMF-STREET          PIC   X%15¤.
00112   001130              07    CMF-CITY-STATE      PIC   X%20¤.
00113   001140              07    CMF-ZIP             PIC   S9%05¤     COMP-3.
00114   001150        05    CMF-SHIPMENT-DATE       PIC   S9%06¤     COMP-3.
00115   001160        05    CMF-TOTAL-ENTRIES       PIC   9%02¤.
00116   001170        03    CMF-ENTRIES.
00117   001180        05    CMF-ORDER-ENTRIES    OCCURS 10 TIMES
00118   001190                            DEPENDING ON CMF-TOTAL-ENTRIES.
00119   001200              07    CMF-STYLE-NO        PIC   S9%05¤     COMP-3.
00120   001210              07    CMF-QTY-ORDERED     PIC   S9%05¤     COMP-3.
00121   001220              07    CMF-BOX-PRICE       PIC   S9%05¤V99  COMP-3.
00122   001230              07    CMF-SUBS-STYLE-NO   PIC   S9%05¤     COMP-3.
00123   001240              07    CMF-QTY-SUBSTITUTED PIC   S9%05¤     COMP-3.
00124   001250
00125   001260 FD  ERROR-LISTING
00126   001270     RECORDING MODE IS F
00127   001280     LABEL RECORDS ARE OMITTED
00128   001290     RECORD CONTAINS 133 CHARACTERS
00129   001300     DATA RECORD IS ERROR-LINE.
00130   001310
00131   001320 01  ERROR-LINE.
00132   001330        03    PRINT-CONTROL            PIC   X%001¤.
00133   001340        03    ERROR-PRINT             PIC   X%132¤.


      4       CRC430              14.25.21        12/01/83


00135   001360 WORKING-STORAGE SECTION.
00136   001370 01  STATUS-VS.
00137   001380        03    S-KEY-1                 PIC   X%01¤.
00138   001390              88    OK-IO                                    VALUE @0@.
00139   001400              88    END-FILE                                 VALUE @1@.
00140   001410              88    INV-KEY                                  VALUE @2@.
```

FIGURE 12-17 *continued*

```
00141   001420        88   PERM-ERR              VALUE @3@.
00142   001430        88   IBM-ERR               VALUE @9@.
00143   001440     03 S-KEY-2                 PIC  X%01¤.
00144   001450        88   SEQ-PASS-ERR          VALUE @1@.
00145   001460        88   DUP-LOGIC-ERR         VALUE @2@.
00146   001470        88   NO-REC-ERR            VALUE @3@.
00147   001480        88   BOUNDARY-ERR          VALUE @4@.
00148   001490        88   FILE-INFO-ERR         VALUE @5@.
00149   001500        88   NO-DLBL               VALUE @6@.
00150   001510
00151   001520 01  VSAM-ERROR.
00152   001530     03 FILLER                  PIC  X%12¤  VALUE @*** CAUSE, @.
00153   001540     03 ERROR-CAUSE             PIC  X%50¤.
00154   001550
00155   001560 01  DATE-TIME-COMPILED.
00156   001570     03 DATE-COMPILE                 PIC  X%08¤.
00157   001580     03 TIME-COMPILE                 PIC  X%08¤.
00158   001590
00159   001600 01  SWITCHES.
00160   001610     03 END-OF-FILE-SW              PIC  9%01¤  VALUE 0.
00161   001620        88   END-OF-FILE                       VALUE 1.
00162   001630     03 REWRITE-SW                  PIC  9%01¤  VALUE 0.
00163   001640        88   REWRITE-CANCELLED                 VALUE 1.
00164   001650     03 RECORD-FOUND-SW             PIC  9%01¤  VALUE 0.
00165   001660        88   RECORD-NOT-FOUND                  VALUE 1.
00166   001670     03 ENTRY-FOUND-SW              PIC  9%01¤  VALUE 0.
00167   001680        88   ENTRY-NOT-FOUND                   VALUE 1.
00168   001690
00169   001700 01  WORK-FIELDS.
00170   001710     03 MF-REC-LENGTH               PIC  9%04¤  COMP.
00171   001720     03 LINE-COUNT                  PIC  9%02¤  VALUE ZEROS.
00172   001730     03 PAGE-COUNT                  PIC  9%04¤  VALUE ZEROS.
00173   001740     03 SUB                         PIC  9%04¤  COMP.
00174   001750     03 SUB1                        PIC  9%04¤  COMP.
00175   001760     03 SUB2                        PIC  9%04¤  COMP.
00176   001770     03 TXS-IN                      PIC  9%09¤  VALUE ZEROS.
00177   001780     03 RECS-UPDATED                PIC  9%09¤  VALUE ZEROS.
00178   001790     03 BAD-TXS                     PIC  9%09¤  VALUE ZEROS.
00179   001800     03 DIS-TXS-IN                  PIC  ZZZ,ZZZ,ZZ9.
00180   001810     03 DIS-RECS-UP                 PIC  ZZZ,ZZZ,ZZ9.
00181   001820     03 DIS-BAD-TXS                 PIC  ZZZ,ZZZ,ZZ9.
00182   001830     03 WORK-COST                   PIC  S9%05¤V99  COMP-3.
00183   001840     03 HOLD-QTY                    PIC  S9%09¤     COMP-3.
00184   001850
00185   001860 01  ERROR-AREA.
00186   001870     03 FILLER                      PIC  X%20¤.
00187   001880     03 ERR-NUMBER                  PIC  9%05¤.


       5      CRC430            14.25.21          12/01/83


00188   001890     03 FILLER                      PIC  X%02¤  VALUE SPACES.
00189   001900     03 ERR-CODE                    PIC  X%03¤.
00190   001910     03 FILLER                      PIC  X%02¤  VALUE SPACES.
00191   001920     03 ERR-MESSAGE                 PIC  X%80¤.
00192   001930     03 FILLER                      PIC  X%20¤  VALUE SPACES.
00193   001940
00194   001950 01  ERROR-HEAD1.
00195   001960     03 FILLER    PIC  X%18¤  VALUE SPACES.
00196   001970     03 FILLER    PIC  X%05¤  VALUE @DATE @.
00197   001980     03 ERR-DATE  PIC  X%08¤.
00198   001990     03 FILLER    PIC  X%20¤  VALUE SPACES.
00199   002000     03 FILLER    PIC  X%26¤  VALUE @C R C 4 3 0    E R R O R    @.
00200   002010     03 FILLER    PIC  X%07¤  VALUE @L I S T@.
00201   002020     03 FILLER    PIC  X%20¤  VALUE SPACES.
00202   002030     03 FILLER    PIC  X%05¤  VALUE @PAGE @.
00203   002040     03 ERR-PAGE  PIC  ZZZ9.
00204   002050     03 FILLER    PIC  X%19¤  VALUE SPACES.
00205   002060
00206   002070 01  ERROR-HEAD2.
00207   002080     03 FILLER    PIC  X%20¤  VALUE SPACES.
00208   002090     03 FILLER    PIC  X%06¤  VALUE @NUMBER@.
00209   002100     03 FILLER    PIC  X%01¤  VALUE SPACES.
00210   002110     03 FILLER    PIC  X%04¤  VALUE @CODE@.
00211   002120     03 FILLER    PIC  X%14¤  VALUE SPACES.
00212   002130     03 FILLER    PIC  X%13¤  VALUE @ERROR MESSAGE@.
00213   002140     03 FILLER    PIC  X%74¤  VALUE SPACES.
00214   002150
```

FIGURE 12-17 *continued*

```
00215    002160 01  RUN-DATE.
00216    002170     03  TODAYS-DATE                    PIC  X%08¤.
00217    002180     03  FILLER                         PIC  X%08¤.


     6          CRC430           14.25.21        12/01/83


00219    002200 PROCEDURE DIVISION.
00220    002210 100-START.
00221    002220     MOVE CURRENT-DATE TO RUN-DATE.
00222    002230     MOVE WHEN-COMPILED TO DATE-TIME-COMPILED.
00223    002240     DISPLAY SPACE.
00224    002250     DISPLAY @DATE COMPILED @, DATE-COMPILE.
00225    002260     DISPLAY @TIME COMPILED @, TIME-COMPILE.
00226    002270
00227    002280     OPEN I-O MASTERFILE.
00228    002290     IF   NOT OK-IO
00229    002300         MOVE @OPEN ERROR@ TO ERROR-CAUSE
00230    002310         PERFORM 999-VSAM-ERROR
00231    002320         STOP RUN.
00232    002330
00233    002340     OPEN INPUT CHANGE-MASTERFILE-TX.
00234    002350     OPEN OUTPUT ERROR-LISTING.
00235    002360
00236    002370     PERFORM 200-READ-TX-FILE
00237    002380         UNTIL
00238    002390             END-OF-FILE.

00240    002410 150-CLOSE-FILES.
00241    002420     PERFORM 500-EOJ.
00242    002430
00243    002440     CLOSE CHANGE-MASTERFILE-TX, ERROR-LISTING.
00244    002450     CLOSE MASTERFILE.
00245    002460     IF   NOT OK-IO
00246    002470         MOVE @CLOSE ERROR@ TO ERROR-CAUSE
00247    002480         PERFORM 999-VSAM-ERROR.
00248    002490
00249    002500     STOP RUN.

00251    002520 200-READ-TX-FILE.
00252    002530     READ CHANGE-MASTERFILE-TX
00253    002540         AT END
00254    002550             MOVE 1 TO END-OF-FILE-SW.
00255    002560
00256    002570     IF   NOT END-OF-FILE
00257    002580         ADD 1 TO TXS-IN
00258    002590         MOVE 0 TO REWRITE-SW
00259    002600         PERFORM 210-READ-MASTER
00260    002610     IF  RECORD-NOT-FOUND
00261    002620         MOVE 0 TO RECORD-FOUND-SW
00262    002630         MOVE @INVALID INVOICE NUMBER@ TO ERR-MESSAGE
00263    002640         PERFORM 300-WRITE-ERROR
00264    002650     ELSE
00265    002660         IF  CMF-NAME-TX
00266    002670             PERFORM 230-NAME-TX
00267    002680             PERFORM 280-REWRITE
00268    002690         ELSE
00269    002700             IF  CMF-ADDR-TX
00270    002710                 PERFORM 240-ADDRESS-TX
00271    002720                 PERFORM 280-REWRITE


     7          CRC430           14.25.21        12/01/83


00272    002730             ELSE
00273    002740                 IF  CMF-NAME-ADDR-TX
00274    002750                     PERFORM 230-NAME-TX
00275    002760                     PERFORM 240-ADDRESS-TX
00276    002770                     PERFORM 280-REWRITE
00277    002780                 ELSE
00278    002790                     IF  CMF-NAME-PHONE-TX
00279    002800                         PERFORM 230-NAME-TX
00280    002810                         PERFORM 280-REWRITE
00281    002820                     ELSE
00282    002830                         IF  CMF-ADDR-PHONE-TX
00283    002840                             PERFORM 240-ADDRESS-TX
00284    002850                             PERFORM 280-REWRITE
```

FIGURE 12-17 *continued*

331

```
00285    002860                        ELSE
00286    002870               IF   CMF-CANCEL-ORDER-TX
00287    002880                       PERFORM 250-CANCEL-ALTER-TX
00288    002890                       PERFORM 280-REWRITE
00289    002900               ELSE
00290    002910               IF   CMF-ALTER-ORDER-TX
00291    002920                       PERFORM 250-CANCEL-ALTER-TX
00292    002930                       PERFORM 280-REWRITE
00293    002940                 ELSE
00294    002950                 IF   CMF-SHIPMENT-DATE-TX
00295    002960                       PERFORM 270-SHIPMENT-DATE-TX
00296    002970                       PERFORM 280-REWRITE
00297    002980                   ELSE
00298    002990                       MOVE @INVALID TX CODE@ TO ERR-MESSAGE
00299    003000                       PERFORM 300-WRITE-ERROR.

00301    003020 210-READ-MASTER.
00302    003030         MOVE CMF-INVOICE-NO TO MF-INVOICE-ORDER-NO.
00303    003040         READ MASTERFILE.
00304    003050         IF  STATUS-VS IS EQUAL TO @23@
00305    003060             MOVE 1 TO RECORD-FOUND-SW
00306    003070         ELSE
00307    003080             IF  NOT OK-IO
00308    003090                 MOVE @READ ERROR@ TO ERROR-CAUSE
00309    003100                 PERFORM 999-VSAM-ERROR
00310    003110                 GO TO 150-CLOSE-FILES.

00312    003130 230-NAME-TX.
00313    003140         IF  CMF-NAME IS NOT EQUAL TO SPACES
00314    003150             MOVE CMF-NAME TO MF-CUSTOMER-NAME
00315    003160         ELSE
00316    003170             MOVE @INVALID NAME TX@ TO ERR-MESSAGE
00317    003180             PERFORM 300-WRITE-ERROR.

00319    003200 240-ADDRESS-TX.
00320    003210         IF  CMF-STREET IS EQUAL TO SPACES
00321    003220             IF  CMF-CITY-STATE IS EQUAL TO SPACES
00322    003230                 IF  CMF-ZIP IS NOT NUMERIC
00323    003240                     MOVE @INVALID ADDRESS TX@ TO ERR-MESSAGE
00324    003250                     PERFORM 300-WRITE-ERROR
```

         8   CRC430   14.25.21   12/01/83

```
00325    003260                     ELSE
00326    003270                     IF  CMF-ZIP IS EQUAL TO ZEROS
00327    003280                         MOVE @INVALID ADDRESS TX@ TO ERR-MESSAGE
00328    003290                         PERFORM 300-WRITE-ERROR.
00329    003300
00330    003310         IF  CMF-STREET IS NOT EQUAL TO SPACES
00331    003320             MOVE CMF-STREET TO MF-CUSTOMER-STREET.
00332    003330
00333    003340         IF  CMF-CITY-STATE IS NOT EQUAL TO SPACES
00334    003350             MOVE CMF-CITY-STATE TO MF-CUSTOMER-CITY-STATE.
00335    003360
00336    003370         IF  CMF-ZIP IS NUMERIC
00337    003380             IF  CMF-ZIP IS NOT EQUAL TO ZERO
00338    003390                 MOVE CMF-ZIP TO MF-CUSTOMER-ZIP.

00340    003410 250-CANCEL-ALTER-TX.
00341    003420         MOVE 0 TO SUB.
00342    003430         PERFORM 265-PROCESS
00343    003440             THRU
00344    003450                 265-EXIT.
00345    003460
00346    003470         MOVE 0 TO SUB2.
00347    003480         MOVE 0 TO WORK-COST.
00348    003490         PERFORM 269-TOTAL-COST MF-TOTAL-ENTRIES TIMES.
00349    003500         MOVE WORK-COST TO MF-TOTAL-COST.

00351    003520 265-PROCESS.
00352    003530         ADD 1 TO SUB.
00353    003540         IF  SUB IS GREATER THAN CMF-TOTAL-ENTRIES
00354    003550             GO TO 265-EXIT.
00355    003560
00356    003570         MOVE 0 TO SUB1.
00357    003580         PERFORM 267-FIND-MF-ENTRY
00358    003590             THRU
```

FIGURE 12-17 *continued*

```
00359    003600              267-EXIT.
00360    003610
00361    003620         IF   ENTRY-NOT-FOUND
00362    003630              MOVE 0 TO ENTRY-FOUND-SW
00363    003640              MOVE @INVALID CANCEL OR ALTER TX@ TO ERR-MESSAGE
00364    003650              PERFORM 300-WRITE-ERROR
00365    003660              GO TO 265-EXIT
00366    003670         ELSE
00367    003680              PERFORM 268-CHANGE-ENTRY
00368    003690                 THRU
00369    003700                    268-EXIT.
00370    003710
00371    003720         GO TO 265-PROCESS.
00372    003730 265-EXIT.
00373    003740         EXIT.

00375    003760 267-FIND-MF-ENTRY.
00376    003770         ADD 1 TO SUB1.
00377    003780         IF   SUB1 IS GREATER THAN MF-TOTAL-ENTRIES
```

```
00378    003790              MOVE 1 TO ENTRY-FOUND-SW
00379    003800              GO TO 267-EXIT.
00380    003810
00381    003820         IF   CMF-STYLE-NO %SUB¤ IS EQUAL TO MF-STYLE-NO-ORDERED %SUB1¤
00382    003830              GO TO 267-EXIT.
00383    003840
00384    003850         GO TO 267-FIND-MF-ENTRY.
00385    003860 267-EXIT.
00386    003870         EXIT.

00388    003890 268-CHANGE-ENTRY.
00389    003900         IF   CMF-CANCEL-ORDER-TX
00390    003910              MOVE 0 TO MF-TOTAL-PRICE %SUB1¤
00391    003920              MOVE @F@ TO MF-ORDER-STATUS %SUB1¤
00392    003930              GO TO 268-EXIT.
00393    003940***       FULL CANCELLATION
00394    003950
00395    003960         COMPUTE HOLD-QTY # CMF-QTY-ORDERED %SUB¤ * +1.
00396    003970***       HOLD-QTY WILL BE THE ABSOLUTE VALUE OF CMF-QTY-ORDERED.
00397    003980
00398    003990         IF   CMF-QTY-ORDERED %SUB¤ IS LESS THAN 0
00399    004000              COMPUTE MF-TOTAL-PRICE %SUB1¤ # MF-TOTAL-PRICE %SUB1¤
00400    004010                   - %HOLD-QTY * CMF-BOX-PRICE %SUB¤¤
00401    004020              SUBTRACT HOLD-QTY FROM MF-QUANTITY-ORDERED %SUB1¤
00402    004030              MOVE @G@ TO MF-ORDER-STATUS %SUB1¤
00403    004040***       ORDER REDUCTION
00404    004050         ELSE
00405    004060              COMPUTE MF-TOTAL-PRICE %SUB1¤ # MF-TOTAL-PRICE %SUB1¤
00406    004070                      + %HOLD-QTY * CMF-BOX-PRICE %SUB¤¤
00407    004080              ADD HOLD-QTY TO MF-QUANTITY-ORDERED %SUB1¤.
00408    004090***       ORDER INCREASE
00409    004100
00410    004110         IF   CMF-SUBS-STYLE-NO %SUB¤ IS NUMERIC
00411    004120         IF   CMF-SUBS-STYLE-NO %SUB¤ IS NOT EQUAL TO ZERO
00412    004130              MOVE CMF-SUBS-STYLE-NO %SUB¤ TO MF-SUBSTITUTED-STYLE-NO
00413    004140                                                          %SUB1¤
00414    004150              MOVE CMF-QTY-SUBSTITUTED %SUB¤ TO MF-QTY-SUBSTITUTED
00415    004160                                                          %SUB1¤
00416    004170              COMPUTE MF-TOTAL-PRICE %SUB1¤ # MF-TOTAL-PRICE %SUB1¤
00417    004180                    + %CMF-QTY-SUBSTITUTED %SUB¤ * CMF-BOX-PRICE %SUB¤¤
00418    004190              IF       CMF-QTY-ORDERED %SUB¤ IS NEGATIVE
00419    004200              AND HOLD-QTY IS EQUAL TO MF-QUANTITY-ORDERED %SUB1¤
00420    004210                   MOVE @B@ TO MF-ORDER-STATUS %SUB1¤
00421    004220***       FULL SUBSTITUTION
00422    004230              ELSE
00423    004240                   MOVE @C@ TO MF-ORDER-STATUS %SUB1¤.
00424    004250***       PARTIAL SUBSTITUTION
00425    004260 268-EXIT.
00426    004270         EXIT.

00428    004290 269-TOTAL-COST.
00429    004300         ADD 1 TO SUB2.
00430    004310         ADD MF-TOTAL-PRICE %SUB2¤ TO WORK-COST.
```

FIGURE 12-17 *continued*

333

```
00432   004330 270-SHIPMENT-DATE-TX.
00433   004340     IF  CMF-SHIPMENT-DATE IS NUMERIC
00434   004350         IF  CMF-SHIPMENT-DATE IS NOT EQUAL TO ZEROS
00435   004360             MOVE CMF-SHIPMENT-DATE TO MF-DATE-SHIPMENT-REQUESTED
00436   004370         ELSE
00437   004380             MOVE @SHIPMENT DATE IS ZERO@ TO ERR-MESSAGE
00438   004390             PERFORM 300-WRITE-ERROR
00439   004400     ELSE
00440   004410         MOVE @SHIPMENT DATE IS NOT NUMERIC@ TO ERR-MESSAGE
00441   004420         PERFORM 300-WRITE-ERROR.

00443   004440 280-REWRITE.
00444   004450     IF  REWRITE-CANCELLED
00445   004460         NEXT SENTENCE
00446   004470     ELSE
00447   004480         PERFORM 290-REWRITE-MASTER.

00449   004500 290-REWRITE-MASTER.
00450   004510     COMPUTE MF-REC-LENGTH # %MF-TOTAL-ENTRIES  *  15¤  +  79.
00451   004520
00452   004530     REWRITE MASTERFILE-RECORD.
00453   004540     IF  NOT OK-IO
00454   004550         MOVE @REWRITE ERROR@ TO ERROR-CAUSE
00455   004560         PERFORM 999-VSAM-ERROR
00456   004570         GO TO 150-CLOSE-FILES.
00457   004580
00458   004590     ADD 1 TO RECS-UPDATED.

00460   004610 300-WRITE-ERROR.
00461   004620     ON  1
00462   004630         MOVE TODAYS-DATE TO ERR-DATE
00463   004640         PERFORM 310-ERROR-HEADINGS.
00464   004650
00465   004660     IF  LINE-COUNT IS GREATER THAN 50
00466   004670         PERFORM 310-ERROR-HEADINGS.
00467   004680
00468   004690     MOVE 1 TO REWRITE-SW.
00469   004700     MOVE CMF-TX-NUMBER TO ERR-NUMBER.
00470   004710     MOVE CMF-TX-CODE TO ERR-CODE.
00471   004720
00472   004730     ADD 1 TO BAD-TXS.
00473   004740     MOVE @0@ TO PRINT-CONTROL.
00474   004750     MOVE ERROR-AREA TO ERROR-PRINT.
00475   004760     WRITE ERROR-LINE AFTER POSITIONING PRINT-CONTROL LINES.
00476   004770     ADD 2 TO LINE-COUNT.

00478   004790 310-ERROR-HEADINGS.
00479   004800     MOVE 0 TO LINE-COUNT.
00480   004810     ADD 1 TO PAGE-COUNT.
00481   004820     MOVE PAGE-COUNT TO ERR-PAGE.
00482   004830
00483   004840     MOVE @1@ TO PRINT-CONTROL.
```

```
00484   004850     MOVE ERROR-HEAD1 TO ERROR-PRINT.
00485   004860     WRITE ERROR-LINE AFTER POSITIONING PRINT-CONTROL LINES.
00486   004870
00487   004880     MOVE @0@ TO PRINT-CONTROL.
00488   004890     MOVE ERROR-HEAD2 TO ERROR-PRINT.
00489   004900     WRITE ERROR-LINE AFTER POSITIONING PRINT-CONTROL LINES.

00491   004920 500-EOJ.
00492   004930     MOVE TXS-IN TO DIS-TXS-IN.
00493   004940     MOVE RECS-UPDATED TO DIS-RECS-UP.
00494   004950     MOVE BAD-TXS TO DIS-BAD-TXS.
00495   004960
00496   004970     DISPLAY SPACE.
00497   004980     DISPLAY @CRC430 CONTROL TOTALS @, TODAYS-DATE.
00498   004990     DISPLAY SPACE.
00499   005000     DISPLAY @INPUT TXS                    @, DIS-TXS-IN.
00500   005010     DISPLAY @MASTERFILE RECORDS UPDATED @, DIS-RECS-UP.
00501   005020     DISPLAY @TXS REJECTED                @, DIS-BAD-TXS.
00502   005030
```

FIGURE 12-17 *continued*

```
00503   005040***  CONTROL TOTALS ARE INTENTIONALLY PRINTED
00504   005050***  ON THE SAME PAGE AS THE JCL.

00506   005070 999-VSAM-ERROR.
00507   005080      DISPLAY @JOB ABORTED@                    UPON CONSOLE.
00508   005090      DISPLAY VSAM-ERROR, @STATUS VS @, STATUS-VS  UPON CONSOLE.
```

FIGURE 12-17 *continued*

and 25 relations called ERR-XX that contin validation error messages corresponding to the 25 forms MEF-01, MEF-02, . . ., MEF-25. The set of these
51 relations constitutes the Form Database for each fiscal year.

A relation MEF-XX consists of the following attributes:

Name	Description	Format
*GID	Grantee ID number	4 digits numeric
*LINE	Line number of a data cell	2 digits numeric
*COL	Column number of a data cell	1 digit numeric
VALUE	Data value appearing in the cell (LINE, COL)	character string or numeric

The attributes marked with an asterisk (*) are key attributes.

Using the CREATE TABLE command in ORACLE, each relation structure is created. Then CREATE UNIQUE INDEX command is used to declare the key attributes. The ORACLE Data Loader Utility is used for loading
data into each relation. Figure 12-19 shows a sample MEF-XX relation.

The above relational structure allows the different attributes VALUE to
be combined with respect to grantee IDs and provide a total of all such
VALUEs. Some of the reports generated by the MEF need such summation
capabilities.

The DB-STATUS relation is used for answering the following questions:

a. Does a particular form relation MEF-XX belong to the Form Database?

b. When was a relation MEF-XX

```
DATE 12/01/83                      C R C 4 3 0   E R R O R

NUMBER CODE                  ERROR MESSAGE

00010  001   INVALID INVOICE NUMBER

00011  002   INVALID ADDRESS TX

00013  004   INVALID NAME TX

00013  004   INVALID ADDRESS TX

00014  007   INVALID INVOICE NUMBER
```

FIGURE 12-18 Error list of CRC 430 run.

GID	LINE	COL	VALUE
4011	02	3	65089
4011	04	3	65089
4011	07	3	65089
4011	08	2	74052
4011	08	3	74126
4011	19	2	88056
4011	19	3	211925

FIGURE 12-19   Sample MEF-XX relation.

i. originally entered

ii. last updated

iii. last processed by the Data Validation subsystem

The relation consists of the following attributes:

Name	Description	Format
*GID	Grantee ID number	4 digits numeric
*FNO	Form Number (e.g., MEF-12)	6 character string
EDATE	Date of first entry of the form into Form Database	DATE
CDATE	Date of last change to any value in the form	DATE
CTIME	Time of the day when the last update was made	4 digits numeric (0000 to 2359)
VDATE	Date of last validation of the form	DATE
FLAG	Integer representing the status of data validation of the form	2 digits numeric

All attributes marked with asterisks (*) are key attributes. Also, DATE in the format column is a built-in data structure in ORACLE that tests for a valid date. For example, the DATE format will not allow a value of September 31, 1983, or February 29, 1982, to be entered into EDATE, CDATE, or VDATE.

Finally, the 25 ERR-XX relations are created corresponding to the 25 forms MEF-XX, where XX ranges from 01 to 25. A relation ERR-XX contains the following attributes:

Name	Description	Format
GID	Grantee ID number	4 digits numeric
CELL	Line and column number of a data cell	3 digits numeric (first 2 digits are line number and the last digit is column number)
ETYPE	Type of validation check (e.g., range, arithmetic, consistency)	5 character string

All the attributes are declared as key.

The Form Database undergoes changes whenever a user enters/edits data via Data Entry/Edit/Display subsystem or when the validation checks are applied via Data Validation subsystem. The relations in the Form Database capture all such changes.

12.4 SUMMARY

This chapter completed the system implementation phase of the two case studies—Toy World, Inc., and Massachusetts Educational Foundation.

For Toy World, the chapter provided the system flowchart, the input/output table, the hierarchy chart, and the program flowchart for the seven programs constituting the order processing subsystem. In addition, the chapter included a system backup plan, a system recovery plan, and the source code listing for the program CRC430 (Update Master File).

For Massachusetts Educational Foundation, the chapter contained the structures of all the 51 relations belonging to the Form Database.

12.5 KEY WORDS

The following key words are used in this chapter:

attribute	program flowchart
data validation	relation
database	system backup plan
hierarchy chart	system flowchart
input/output table	system recovery plan
key attribute	

REVIEW QUESTIONS

1. For the programs CRC410 and CRC440, prepare complete documentation consisting of
   a. Structure chart
   b. Pseudocode
   c. Source code listing in COBOL
2. Using dummy data to create the necessary input files, run the program CRC430 to generate an output file.
3. What kind of user training do you recommend for the order processing subsystem for Toy World?
4. Prepare a system backup plan for the financial reporting system for MEF.
5. Suppose that MEF wants to create a single Form Database that will contain *all* the relations for *all* fiscal years. How would you change the structures of the 51 relations to accommodate this change?

FORMULATIONAL PROBLEM

Refer to the structure of the Form Database used by the financial reporting system that is being developed for the MEF. Assume the following scenario:

a. On August 15, 1983, at 10:20 A.M. user XYZ logs in to enter data *for the first time* from Form MEF-04 for grantee ID 4008.
b. On August 15, 1983, at 8:00 P.M. validation checks are applied to the above form data in the Forms Database.
c. On August 19, 1983, at 2:05 P.M. user XYZ logs in to reenter/edit data for the same form in order to correct the validation errors reported in step (b).
d. Same as step (b) but occurring on August 19, 1983, at 8:00 P.M. No validation errors are detected so that the cycle ends here.

Analyze each of the above steps to determine the changes that take place at each one. Use fictitious data in the form MEF-04 (see Figure 8-29 for the format of this form), and display the contents of the three relations

```
MEF-04
DB-STATUS
ERR-04
```

at the end of each step.

CONTEMPORARY ISSUES IN SYSTEM DEVELOPMENT

Part V consists of Chapters 13 through 15. Chapter 13 gives an overview of a decision support system, its impact on an organization, and its relationship to the MIS. Chapter 14 is a brief description of database systems. Chapter 15 discusses seven heterogeneous topics that have not been covered elsewhere in the book.

13

Decision Support Systems in an Organization

13.1 ROLE OF DECISION SUPPORT SYSTEMS

In Section 1.3 we discussed the functions of three levels of management. The top and the middle levels deal with semistructured and unstructured tasks that involve decision making under uncertain circumstances. A decision support system, DSS for short, can help them in this situation. It has the following objectives:

a. It helps managers at the upper levels.
b. It is flexible and responds quickly to managers' questions.
c. It provides "what if" scenarios.
d. It takes into account the personal decision-making styles of managers.

In order to meet these objectives a DSS should have the following characteristics:

a. Relies heavily on sophisticated quantitative techniques of model building
b. In cases in which an analytic optimizing model cannot be solved, relies on simulation
c. Uses statistical analysis to collect data and predict trends

This chapter is based on Mittra ([2], Chapter 1).

d. Uses features whereby noncomputer-oriented people can use it in an interactive mode
e. Is usually aimed at semistructured and unstructured tasks
f. Is designed to remain flexible and adaptible so that it can be modified so as to meet the specific decision-making style of the user

A semistructured or unstructured task involves a decision process that is partly routine and partly judgmental. The routine part can be easily automated, but the judgmental part has to be developed by a manager. A DSS is developed mainly by working from a manager's perspective and accepting his or her implicit definition as to which components should be left to personal judgment. It provides a delicate balance between human judgment and automated procedure. It adopts a coherent strategy for going beyond the traditional use of computers in fairly structured situations while avoiding ineffectual efforts to automate inherently unstructured ones. The DSS thus thrives best on semistructured tasks.

The above observation leads us to the requirement that a DSS should be an adaptive system. It adapts itself to changes of several kinds over three time horizons. Sprague and Carlson ([3], Chapter 1) comment as follows:

> In the short run, the system allows *search* for answers within a relatively narrow scope. In the intermediate time horizon, the system *learns* by modifying its capabilities and activities (the scope or domain changes). In the long run, the system *evolves* to accommodate much different behavior styles and capabilities.

13.2 TECHNICAL CAPABILITIES OF DSS

We are now ready to attempt a formal definition of DSS. A DSS can be described as a computer-based information system that helps a manager make key decisions and thereby improves the effectiveness of the manager's problem-solving process. Like any other information system, a DSS consists of both hardware and software. However, a DSS always contains a feedback loop, which helps managers answer what if questions. As a result, the DSS has to be implemented as an on-line interactive system whereby the user can easily begin and be guided by the system, preferably in a menu-driven manner. The DSS provides adequate HELP routines so as to enable its user to select paths. This is normally described by saying that the DSS should always have a user-friendly front-end since it is geared primarily toward users who are less computer oriented. A typical feedback loop in a DSS can be depicted as in Figure 13-1.

We can summarize a situation in which DSS can be useful as involving some or all of the following characteristics:

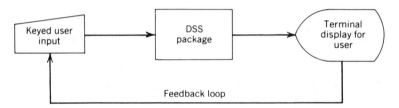

FIGURE 13-1 Feedback loop. (Reproduced with permission from S.S. Mittra, *Decision Support Systems*, Wiley, New York, 1986.)

a. Existence of a database so large that the manager has difficulty accessing and making conceptual use of it
b. Necessity of manipulation or computation in the process of arriving at a solution
c. Existence of some time pressure for the final answer
d. Necessity of judgment to decide upon available alternatives by asking lots of what if questions

The DSS requires a unique approach to systems analysis and design. The usual process of designing and implementing a system has to be interfaced with rapid and frequent user feedback to ensure that the DSS being built will ultimately address the decision-making needs of the managers.

The first state in DSS development is decision analysis, with the manager defining the key decision problems. The manager can best recognize, with the system designer's help, the particular aspects that can be improved and the components that have the most overall impact on the effectiveness of decision making. There are four levels of support available from a DSS ([1], Chapter 4):

a. Access to facts or information retrieval. For a manager to find relevant information in a mountain of raw data can be a nontrivial job.
b. Addition of filters and pattern recognition ability to look for information selectively and give conceptual meaning to data.
c. Capability of simple computations, comparisons, and projections.
d. Development of useful models for the manager. The model must be so designed as to provide managers with answers they can and will act on.

The design and implementation of a DSS require sophisticated mathematical modeling techniques (e.g., linear programming and statistical forecasting), simulation methods, and high-powered computer support. In the past, traditional mathematical models and MIS have generally not allowed such interaction because of the following reasons:

1. Gap between managers and MIS specialists
2. High degree of specialization of the computer world
3. Nature of decision making not amenable to computerization

But recent changes in technology have made the implementation of such systems possible.

13.3 OPERATIONAL SCENARIO FOR A DSS

We now present a hypothetical session involving a DSS. Suppose that you, as the director of personnel in your organization, have been asked by the president of the company to prepare a projection of staffing needs for the next 5 years. The company has a well-developed DSS supported by a sophisticated DBMS, which provides a query language and graphics capability. We can visualize the DSS in operation as follows:

1. You begin as an authorized user via a CRT terminal with graphics capability.
2. The DSS displays a menu listing a set of options from which you have to select one. You can also call for a HELP file or can EXIT.
3. You select the option STAFFING NEEDS.
4. The DSS provides a series of prompts by which it collects data from you in order to perform the projection. Thus, you communicate to the system the time frame, (i.e., 5 years), the frequency of projection (i.e., projection on annual basis), the different job classifications, the salary levels, and so on.
5. The DSS then searches the personnel database in order to gather past historical data on staffing. It then feeds these data to a statistical projection module to determine the annual staffing needs for the next 5 years. Finally, it displays at your terminal this data in a report format.
6. If you are satisfied with this ad hoc report, you can get a printed hardcopy of it. Otherwise, you may want to experiment with the projected data by changing your input. Herein is the typical what if capability of a DSS.
7. At the end, you can get a graphics display of the projected data and obtain a hardcopy of the graph.

Figure 13-2 illustrates our situation graphically.

The above scenario, though simple enough, embodies the essential features of a DSS session. In order to make this session possible, the designers of the DSS have to use a set of sophisticated quantitative tools. We next formulate a list of such tools, with a brief description and justification of each.

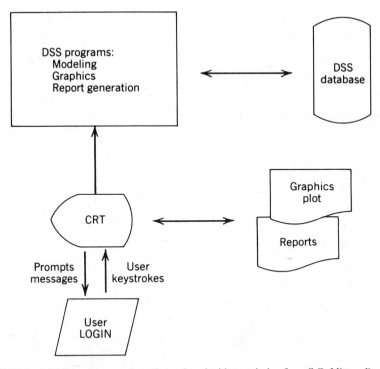

FIGURE 13-2 DSS interactive session. (Reproduced with permission from S.S. Mittra, *Decision Support Systems*, Wiley, New York, 1986.)

13.4 TOOLS FOR BUILDING A DSS

Conceptually, a DSS consists of the following four modules:

a. Control module
b. Data storage module
c. Data manipulation module
d. Model building module

The *control module* (CM), also known as the DSS front-end, is usually menu driven and interfaces with the other three modules. It provides prompts and messages to guide users in formulating their problems and generating the responses. It accepts user keystrokes and interprets them as specific action initiating commands. The CM has to be user friendly in the sense that it can help a noncomputer-oriented user with the DSS. The prompts and messages should be written in plain English as far as practicable and should avoid computer jargon.

The *data storage module* (DSM) contains all the data required by the DSS. It is really the complete database for the organization. The DSM should preferably be structured as a relational database instead of a network model, because generation of reports is comparatively simple when the database uses a relational model. The DSM provides all the current data as well as past historical data required by the model building module.

The *data manipulation module* (DMM) is responsible for retrieving data from the DSM and producing reports and/or graphs using these data. The DMM is supported by a nonprocedural query language, a report generating package, and a graphics package. Normally, the DBMS that supports the DSM provides all these features. As a minimum, the graphics includes business tools such as bar graphs, pie charts, and plots.

The *model building module* (MBM) contains all the quantitative model building software. It uses optimizing modeling principles, statistical analysis, forecasting algorithms, decision analysis methods, and simulation principles. Usually the MBM is the most difficult component in a DSS since the selection of appropriate algorithms from the vast repertoire of mathematical models is a nontrivial job. In addition, if a specific modeling algorithm does not lead to an explicit solution, recourse has to be taken to the simulation method. Figure 13-3 is a schematic of the four component modules of a DSS.

Since the control module provides interfaces to all the other modules, it

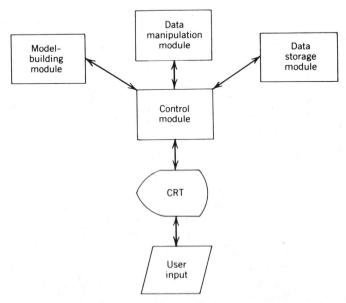

FIGURE 13-3 Component modules of a DSS. (Reproduced with permission from S.S. Mittra, *Decision Support Systems*, Wiley, New York, 1986.)

should be designed and implemented first. The success of a DSS depends to a large extent on the efficiency of its control module.

13.5 IMPACT OF A DSS ON MANAGEMENT

A DSS is designed and implemented with managers as the end users. Its impact is on decisions for which there are sufficient structure for computer tools and analytical techniques to be useful but where the manager's judgment is essential. The real payoff from the manager's viewpoint is in extending the range and capability of his or her decision process to make it more effective. The DSS appears to a manager as a supportive tool under his or her own control, which does not attempt to automate the decision process, predefine objectives, or impose solutions. It accepts input from the manager, processes it, and then provides the output for review. If the output is not satisfactory, the manager can repeat the process until the solution is satisfactory, as shown in Figure 13-1.

It must be clearly understood that no DSS can ever take the decision-making authority away from managers. Many of the managers, especially those who are not familiar with computer technology, have an inherent fear of the DSS. They fail to realize, however, that DSS is just a tool to be used for making better and quicker decisions.

Whereas the managers play the role of end users of a DSS, the analysts and the designers get involved with the actual building of the system. The latter two groups are composed of technical people who mostly belong to the middle management level in a company or are outside consultants from vendors of DSS. The analysts work as technical liaison persons between managers and designers. They must have some familiarity with the general problem area and be comfortable with the information system technology components and capabilities. The designers are the really technical people who develop new analytical models, new databases to store data for easy access and retrieval, and appropriate prompts and display format for user friendly interaction with the DSS. In this task, the designers use or even develop new hardware and software.

13.6 SUMMARY

This chapter described what a decision support system (DSS) is and how it impacts on an organization.

The DSS appears as a subsystem within the MIS. Its aim is to help managers address semistructured and unstructured tasks, because the structured tasks can easily be handled by the application subsystems. In order to meet its goal, the DSS has to be designed and implemented as an interactive system with a user friendly front end. The prompts and messages given by this front end should be in plain English as far as practicable and should avoid computer jargon because the users of a DSS are not always computer oriented. The main requirement for a DSS is to enable its users to experiment with what if scenarios. A manager always likes to examine alternatives before coming to a decision, especially when the decision has far reaching consequences.

After describing a hypothetical operational scenario illustrating the use of DSS in action, the chapter provided a list of technical requirements for a DSS. Conceptually, a DSS consists of four components or modules:

☐ Control module
☐ Data storage module
☐ Data manipulation module
☐ Model building module

Each module has been described briefly regarding its function and justification.

13.7 KEY WORDS

The following key words are used in this chapter:

decision-making authority	strategic level of management
Decision Support System (DSS)	tactical level of management
Management Information System (MIS)	time trend
	unstructured task
model building	user friendly front end
query language	what if question/scenario
semistructured task	
simulation	

REFERENCES

1. P. G. W. Keen and M. S. Scott Morton, *Decision Support Systems: An Organizational Perspective*, Addison-Wesley, Reading, MA, 1978.
2. S. S. Mittra, *Decision Support Systems: Tools and Techniques*, Wiley, New York, 1986.
3. R. H. Sprague and E. D. Carlson, *Building Effective Decision Support Systems*, Prentice-Hall, Englewood-Cliffs, NJ, 1982.

REVIEW QUESTIONS

1. Distinguish between a structured, a semistructured, and an unstructured task. Why is a DSS more effective for the latter two types?

2. Explain the statement, The structured tasks can be handled by the application subsystems of the MIS and do not need a DSS.

3. List the technical capabilities of a DSS.

4. Why do you need mathematical modeling and simulation for building a DSS?

5. Identify the four modules of a DSS, and describe the function of each.

6. Do you think that a DSS can take away some decision making authority away from managers? Explain your answer fully.

7. How do you measure the success or failure of a DSS? (The answer is *not* in the book.)

14

Impact of Database Systems on the System Development Process

14.1 NEED FOR DATABASE SYSTEMS

During the present decade online systems are becoming progressively popular. The advent of microcomputers has significantly contributed to this trend. The end users, especially the middle to top level managers, are now equipped with microcomputers and want to generate ad hoc reports in response to daily information requirements. A file processing environment is not conducive to this situation for a variety of reasons of which the two main ones are

   a. Normally a separate program is required for each report.
   b. Data redundancy is widespread among files related to the same application.

   Ideally, an organization should strive for a centralized data repository with little or no redundancy of data. Software should exist that can access these data and produce ad hoc reports without having the end user write complex programs. The database system provides such an environment. In fact, one of the early expectations of organizations establishing computer-based information systems was that the systems would be able to integrate all organizational data into a single all-encompassing database to serve management's information requirements. When organizational data are not integrated, redundancy in data collection, processing, storage, and report generation occurs.

We now discuss some basic features of a database system. For more detailed information, refer to reference 2 or 3.

14.2   FILE PROCESSING VERSUS DATABASE SYSTEM

Database technique allows an organization's data to be processed as an integrated whole. It thereby reduces the fragmentation imposed by separate files designed and implemented for separate applications and enables users to access data from different parts of the database to suit their purposes. File processing systems are indeed predecessors of database systems.

Consider the situation shown in Figure 14-1. Three separate files are handled by three separate application programs (Customer Listing, Invoice, Inventory) to print three separate reports. Each program interfaces separately with the operating system during this process to generate the reports.

Let us next look at Figure 14-2. All the data needed by the three programs are stored in a single logical file or schema. The same reports that were generated using separate files can now be produced through small application programs using DML (Data Manipulation Language) commands and statements from a host language like FORTRAN, BASIC, or Pascal or by using QUERY language. These programs interface with complex software called Data Base Management System (DBMS), which, in turn, interfaces with the operating system to generate the necessary reports.

A file processing system, although an efficient data management tool for a specific application, has the following disadvantages:

a. **Redundancy of data.** Identical data are distributed over various files (e.g., Customer Number appears in Customer File and Invoice File).

b. **Multiple update.** One field may be updated in one file but not in others, thereby leading to lack of data integrity and production of conflicting reports.

c. **Waste of storage space.** When the same field is stored in four different files, the required storage space is needlessly quadrupled.

d. Access language normally unique to applications programmer and often not user friendly.

A database system reduces or eliminates these problems; its advantages are

a. Reduction of data redundancy

b. Maintenance of data integrity

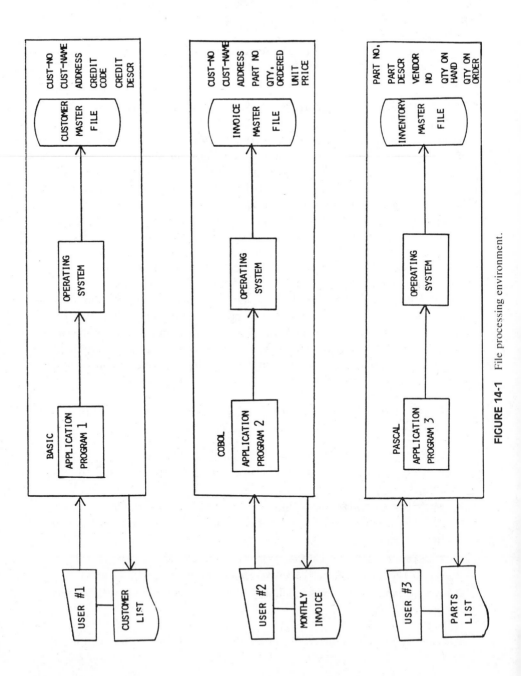

FIGURE 14-1 File processing environment.

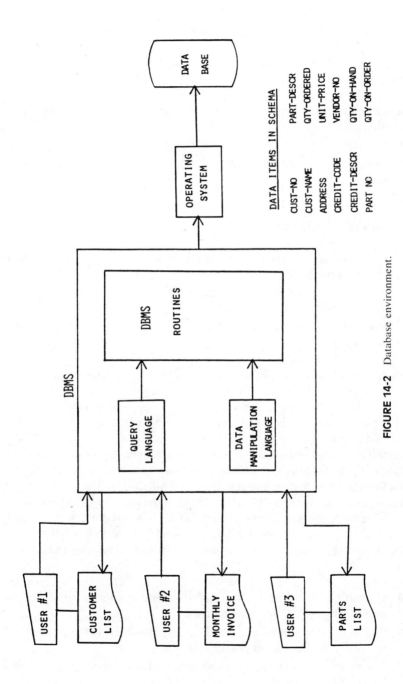

FIGURE 14-2 Database environment.

DATA ITEMS IN SCHEMA

CUST-NO	PART-DESCR
CUST-NAME	QTY-ORDERED
ADDRESS	UNIT-PRICE
CREDIT-CODE	VENDOR-NO
CREDIT-DESCR	QTY-ON-HAND
PART NO	QTY-ON-ORDER

   c. Reduction in wasted storage space

   d. Access to data through user friendly language

However, a database system is not an unmixed blessing. Some of its disadvantages are

   a. Installation and maintenance are more expensive than that of separate file systems.

   b. Backup and recovery are more difficult.

   c. Failure of one component of an integrated system may stop the entire system.

14.3   DATA MODELS

The logical structure or schema of a database can be described by using one of three data models:

   ☐ Hierarchical

   ☐ Network

   ☐ Relational

The hierarchical model was introduced in the late 1960s. This model structures data in a rigid owner/member (also called parent/child) relationship. Any specific data item in the schema can be accessed only by following a path from the owner to the member record. In the early 1970s, the Conference on Data Systems Languages (CODASYL) Data Base Task Group (DBTG) created the network model as an improvement over the hierarchical model. In the network model, a specific data item can be accessed directly from the control module by creating appropriate paths called sets. The relational model was originally introduced by Dr. E. F. Codd in the early 1970s. Many database professionals think that the relational model is *the* natural way to look at data and that this model will be the most widely used in the future.

   The following comments can be made by way of comparing the three data models:

   a. Hierarchical and network data structures require a predetermined structuring of data links or paths.

   b. Those two structures are good for repetitive reports or formats but are less efficient for ad hoc queries.

   c. Relational data structure is excellent for ad hoc queries but less efficient for repetitive reports.

   In a relational database, different files exist as separate relations and, therefore, there is no structure like owner/member relationship among these relations. The schema structure is not predetermined, and relations can be added or deleted from the schema without affecting the other relations.

This flexibility of schema modification makes the relational database suitable for relatively unstructured exploratory applications. Once we determine the precise structure of the data that is needed for producing the necessary reports, the relational database can be converted into a network database.

In 1976 Chen [1] introduced a new data model called entity-relationship model. It is a hybrid between network and relational models. It consists of two types of relations, entity and relationship. An entity relation is the same as the relation in a relational model.

A relationship relation is the relational version of the set construct of a network model. As of 1986, this model is not implemented by any commercially available package. However, a relational database management system can implement an entity-relationship data model.

14.4 DATABASE MANAGEMENT SYSTEM

A database management system (DBMS) is software that handles all accesses to a database. It works in conjunction with an operating system and provides the following capabilities for the user, who can be an application programmer or a nonprogramming end user:

a. Creates the structure of each record according to the data model selected.

b. Creates the correspondence (called set in CODASYL terminology) between two record types in a hierarchical or network data model.

c. Loads data into the database.

d. Updates (i.e., add, delete, or change) records in the database.

e. Generates reports from the data in the database.

f. Maintains integrity of data in the database.

g. Maintains data security by means of multilevel passwords. For example, a password can be used to control access to the entire database, or to selected records within the database, or to selected data items within a record.

The data definition (or description) language (DDL) of a DBMS handles items (a) and (b). The data manipulation language (DML) is responsible for items (c) through (e). Thus, DML handles two types of functions: retrieving data and updating a record. The former can also be done by the query language or the report writer or the host language interface. A query language is a nonprocedural fourth-generation language that can be used to generate a report simply by specifying what is needed in the report. Once the contents of a report are finalized, its format can be improved by using a report writer. This latter software allows users to print titles, footnotes, page numbers, dates, and subtotals in a report. If a report is fairly complex, a query language may not be able to produce it. In that case, a host language interface can be used. Such an interface works in conjunction with the DBMS and any high-level language such as FORTRAN, COBOL, or Pascal. The user writes a program in the appropriate language and combines it with host language interface commands to generate the report. In addition to the host language interface, some DBMS packages provide a separate procedural language of its own for producing complex reports. Optionally, a graphics package may also be provided by the DBMS to produce graphics such as bar charts, pie charts, and graphs from the data in the database.

The data dictionary of the DBMS maintains the data integrity in the database. It is often called a "database about the database." It contains information on the record format, key data items, owner and member of each set in a hierarchical or network database, access previleges of users, and so on. The password protection scheme of the DBMS is responsible for data security in the database.

To sum up, a DBMS consists of the following components:

☐ DDL
☐ DML
☐ Query language
☐ Report writer
☐ Graphics generator
☐ Host language interface
☐ Procedural language of its own

The query language and the report writer are often regarded as parts of the DML. Also, DDL and DML are together called the data sublanguage, or DSL (see [2], pp. 32–35).

14.5 RELATIONAL DATABASE

In a relational database the data are stored as separate *relations*. A relation is a two-dimensional table with the following properties:

a. Entries in the table are single valued; i.e., neither repeated groups nor arrays are allowed.
b. Entries in any column are all of the same type.
c. No two rows in the table are identical.
d. Order of rows or columns is immaterial.

Each column of a relation is called an *attribute,* and each row is called a *tuple.* If the relation has *n* attributes, then each of its rows is an *n*-tuple.

One or more of the attributes in a relation can be declared as a *key* for the relation. The key attribute(s) must be so designed as to uniquely determine a row when a value for the key is given. This specification creates a separate index file for the relation by using the key as the index field. The index processing greatly improves the access time and the system efficiency for queries from users by retrieving rows from a relation where an attribute has a specific value.

14.6 NORMALIZATION OF RELATIONS

A relation satisfying conditions (a) through (d) of Section 14.5 is said to be in *first normal form*. This is the most general type of relation. A relation is said to be in *second normal form* if it is in first normal form and one of the following condition applies:

1. Key consists of a single attribute.
2. No nonkey attribute exists.
3. Every nonkey attribute depends on the entire key.

The above definition brings us to the issue of dependency among attributes. An attribute A is *functionally dependent* on attribute B if, irrespective of any additions or deletions of rows at any time, the value of B determines the value of A. In a relation, any attribute must be functionally dependent on the key attribute(s), if any, for the relation. If, however, there is an additional dependency among nonkey attributes that does not involve the key, there is said to be a *transitive dependency* among the attributes.

A relation is said to be in *third normal form* if it is in second normal form and if there is no transitive dependency among its attributes.

As an example, refer to the Customer Master File of Figure 14.1. We can design it as a relation with the following attributes:

☐ Customer Number
☐ Customer Name
☐ Customer Address
☐ Credit Code
☐ Credit Description

This satisfies conditions (a)–(d) of Section 14.5 and hence is in first normal form. Using the customer number field as a key, we can see that this is also in second normal form. But since the credit code determines the credit description and since this dependency does not involve the key attribute, customer number, there is a transitive dependency among the attributes. Hence the relation is *not* in third normal form. However, it can be split into the following two relations, each in third normal form:

Customer Relation A	Customer Relation B
Customer Number	Credit Code
Customer Name	Credit Description
Customer Address	
Credit Code	

It should be pointed out here that there is nothing sacred about a relation being in third normal form. Usually a relation in first or second normal form exhibits certain anomalies with respect to addition, deletion, or update of attributes, whereas a relation in third normal form does not have such problems. Consequently, a relation in third normal form is preferred. However, any relation can be divided into relations of higher normal forms.

To illustrate the difficulty encountered with first or second normal forms, suppose the customer relation has the following values:

Customer Number	Customer Name	Customer Address	Credit Code	Credit Description
1	Smith	Boston	A2	Very good
2	Jones	Allentown	B	Good
3	Brown	New York	X	Not determined
4	Adams	Boston	B	Good
5	Blake	Cambridge	A2	Very good

If the row for customer 3 is deleted, the information about credit code X is also gone. However, by splitting g the customer relation into two separate relations, A and B, the data on customer 3 will be deleted from relation A (which is the file on current customer), but relation B (which is a master list of all possible credit codes and descriptions) will still contain data on credit code X.

14.7 RELATIONAL DBMS

In a relational database the data are modeled in the natural form of relations, which are processed by performing logical operations on attributes. These operations are expressed in terms of relational algebra. All such operations result in a collection of rows with attributes and values. Some of these collections are new relations created by the relational algebra commands. Any database management system that can access a relational database and generate responses to ad hoc queries and reports by using relational algebra is called a relational DBMS.

A relational DBMS always contains a set of basic commands that allows the manipulation of data by the query language of the DBMS to generate ad hoc reports. The following operations can be regarded as one such set of basic commands:

a. SELECT. Retrieves data from a relation based on logic conditions

b. PROJECT. Creates new relations as a subset of existing relation

c. INTERSECT. Creates new relation from two relations based on common values of specified attributes

d. JOIN. Creates new relation from two relations based on one attribute in each relation against a logic condition

Although the terminology used here is taken from set theory, no knowledge of set theory or mathematical logic is needed to use these commands to find answers for ad hoc queries.

14.8 NONRELATIONAL DATABASE

Date has described three representative nonrelational systems: an inverted list system, a hierarchical system, and a network system ([2], Chapters 21–23). In this section we give an overview of a network database as s an example of a nonrelational database.

A network database consists of two types of entities: a record type and a set type. A record type is the same as a record in a file. It consists of a number of fields, each of which has a specific format (e.g., integer, character string) and contains some data value. A set type is a link between two distinct record types, one of which is the owner and the other is the member of the set. A set can be of one-many or many-one or many-many type. A one-many set has one owner and zero or more members; a many-one set has one or more owners and zero or one member; a many-many set has one or more owners and zero or more members.

A set type is implemented by a chain of pointers that originates at the owner occurrence, runs through all the member occurrences, and finally returns to the owner occurrence. The implementation of a set type can be done by the linked list data structure. Any other implementation method is functionally equivalent to the chain of pointers method.

The network data model, as initially created by the CODASYL DBTG report (see Section 14.3), did not support a many-many set type. However, some vendors extended the capability of a network model by including the many-many set type.

The query language of a nonrelational DBMS is always procedural in nature. This means that the user must specify in the request both *what* is needed for the report as well as *how* the relevant data are to be retrieved from the database. Any optimization in the data retrieval process is done by the user and not by the system. Therefore, such a query language is more suitable for an application programmer as opposed to an end user.

14.9   EXAMPLE OF A DATABASE

The database discussed here is called SAMPLE. It concerns a company with several departments, each department having several employees, and each employee having a number of skills and a number of dependents. . The schema for SAMPLE consists of the following four relations:

1. Relation: DEPMNT

Attribute Name	Type	Length	Key
DEPTNO	INT	3	Yes
NAME	TEXT	20 characters	
MGR	TEXT	20 characters	

2. Relation: EMPLOYEE

Attribute Name	Type	Length	Key
DEPTNO	INT	3	
EMPNAM	TEXT	20 characters	
ADDR	TEXT	50 characters	
IDNO	INT	4	Yes
SKLCOD	INT	2	

3. Relation: SKILL

Attribute Name	Type	Length	Key
SKLCOD	INT	2	Yes
SKLNAM	TEXT	20 characters	

4. Relation: DEPNDENT

Attribute Name	Type	Length	Key
IDNO	INT	4	Yes
DEPNAM	TEXT	20 characters	
DEPAGE	INT	2	
DEPSEX	TEXT	1 character	

As is seen, some duplication of data items exists in our database. This is true, in general, for any relational database due to the lack of structure among the relations. In order to respond to ad hoc queries involving two relations, it is necessary to have duplicate attributes in those two relations to connect them.

We now give the network version of the same data base. Figure 14-3 shows the schema structure of the database. It contains four record types and three set types as listed below:

Record Types

---

1. DEPMNT (DEPTNO, NAME, MGR)

2. EMPLOYEE (EMPNAM, ADDR, IDNO)

3. SKILL (SKLCOD, SKLNAM)

4. DEPNDENT (DEPNAM, DEPAGE, DEPSEX)

The fields of a record are listed within parentheses after that record. Their types and lengths are the same as in the relational version:

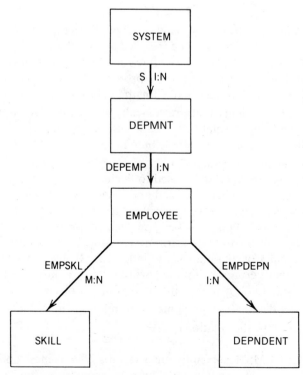

FIGURE 14-3 Network schema for SAMPLE database.

Set Types

Name	Type	Owner	Member
DEPEMP	1:N	DEPMNT	EMPLOYEE
EMPSKL	M:N	EMPLOYEE	SKILL
EMPDEPN	1:N	EMPLOYEE	DEPNDENT

The notations 1:N and M:N represent, respectively, one-many and many-many correspondences. Also, the set type S in Figure 14-3 is SYSTEM owned and is often called a singular set. It should be noted that the data redundancies in the relational version are removed in the network version because of the set types.

14.10   DESIGN AND IMPLEMENTATION OF A RELATIONAL DATABASE

The following checklist can be used to design and implement a database once the decision has been made that a database system is needed for an organization:

a. Gather data requirements. This involves listing all fields, with their descriptions and formats, to be included in the schema. This is primarily a requirements analysis phase. Each field must be semantically analyzed and understood before it is properly defined. The end product of this step is a data dictionary for the database.

b. Combine those fields that are logically related into separate groups, each group being a tentative relation. This is a prerequisite for designing the individual relations and attributes.

c. For each relation decide on a key field, if needed. Find if the DBMS has a command that can be used to declare certain attribute(s) as key after a relation has been defined and loaded. We recommend that for large relations (say, 1000 or more rows) it is most efficient *not* to specify a key at the definition stage because such specification causes an index file to be built for the key attribute resulting in the additional cost of building and storing data for the index file. If, however, it is known for sure that certain attribute(s) will be used as key, then such declaration should be made during the definition phase.

d. Create the database to implement the relations in the schema.

e. Load data into the relations.

The use of a DBMS significantly enhances the effectiveness and efficiency of the information systems developed by system analysts. It makes it easier to integrate the collection, processing, storing, and reporting of organizational data and also enhances the involvement of top management in the system.

14.11 SUMMARY

Recognizing the fact that databases are becoming widely popular, this chapter provided a brief overview of database concepts. It compared a file processing system with a database system and pointed out the relative advantages of the latter. A database can be designed by using any one of three data models: hierarchical, network, or relational. A brief mention was made of the entity-relationship model introduced by Chen in 1976 ([1]).

A DBMS is the tool used for implementing a database design. It consists of the following components:

- ☐ Data definition language (DDL)
- ☐ Data manipulation language (DML)
- ☐ Query language
- ☐ Report writer
- ☐ Graphics generator
- ☐ Host language interface
- ☐ Procedural language of its own (optional)

The chapter then discussed relational databases and the normalization process of relations. A relational DBMS uses the four basic operations of SELECT, PROJECT, JOIN, and INTERSECT to manipulate data in a relational database. Network database was then introduced as an example of a nonrelational database. An example database, called SAMPLE, was discussed in both relational and network versions in order to illustrate the concepts introduced so far.

The chapter gave a checklist of items for design and implementation of a relational database. It closed with comments on system analysis activities in a database environment.

14.12 KEY WORDS

The following keywords are used in this chapter:

ad hoc report	database management system
application programmer	(DBMS)
attribute	data definition language (DDL)
database	data manipulation language (DML)

data redundancy
file processing
first normal form
functionally dependent
graphics generator
hierarchical data model
host language
host language interface
key
member of a set
multiple update
network data model
nonprocedural language
normalization

owner of a set
procedural language
query
query language
record type
relation
relational database management system
relational database
report writer
second normal form
set type
third normal form
transitive dependency

REFERENCES

1. P. P.-S. Chen, *The Entity-Relationship Model—Toward a Unified View of Data*, ACM Transactions on Database Systems, vol. 1, no. 1, March 1976, pp. 9–36.
2. C. J. Date, *An Introduction to Database Systems*, vol. 1, 4th edition, Addison-Wesley, Reading, MA, 1986.
3. David Kroenke, *Database Processing*, SRA, Chicago, 1977.

REVIEW QUESTIONS

1. Explain why database technology is needed for today's information requirements. Illustrate with an example from your own experience.
2. List the advantages and disadvantages of a database system over a file processing system.
3. Define the three data models used in a database.
4. What are the primary functions of a DBMS? List the components of a DBMS that perform these functions.
5. What is a nonprocedural language? Why should a query language be nonprocedural?
6. How do you distinguish between a host language interface of a DBMS and a procedural language that is unique to the DBMS? (The answer is not in the chapter.)
7. Define these terms: first normal form, second normal form, and third normal form.

8. What is an unnormalized relation? Give an example of such a relation, and convert it to a first normal form.

9. Define transitive dependency and give an example.

10. What are four basic operations of a relational DBMS?

11. Explain and illustrate the set type in a network database.

12. What are the organizational and operational implications of performing system analysis in a database environment?

13. How does a DBMS make it easier to change the structure of a database?

14. Do you think that a database environment is beneficial for online applications alone? How does it affect a batch processing system?

15. What problems may arise if an organization uses two or more DBMS packages from different vendors? (The answer is not in the book).

Miscellaneous Topics

15.1 OBJECTIVES OF THE CHAPTER

This chapter discusses seven heterogeneous topics pertinent to the system development process that have not been addressed in the previous chapters. They are the following:

1. Information center concepts
2. Interaction among managers, MIS staff, and computer operations staff
3. Consultant help versus in-house expertise
4. Third-party leasing of equipment
5. New methodology for structured development
6. On-line transaction processing
7. System development for expert systems

15.2 INFORMATION CENTER CONCEPT

Alan Freedman, in *The Computer Glossary,* has defined an information center as follows:

> A department dedicated to user-oriented computing; it is a separate section within the MIS department which provides tools, assistance and training to users for user-oriented and user-controlled computer activities. Products like query languages, report writers and financial planning systems are examples

of the software packages which are provided. Information center consultants work with users to make available the necessary files and databases from the main computer to assist users.

Despite this definition, an information center means different things to different people. To some, it may be a workstation; to others, a large database; and to still others, a separate room equipped with terminals and software. Many people even think that information center is just another buzzword like decision support system or office automation or user friendly.

The information center was first introduced by IBM and heavily promoted during the early 1980s, although the concept has been around since 1975. The overall industry reaction toward information center has been positive. In a 1983 survey by FTP, a market research group in Port Jefferson, NY, it was found that 63 percent of the 71 user organizations surveyed had already implemented a center and an additional 24 percent planned to implement one within the next 2 years. Of the 63 percent organizations that implemented centers, 38 percent of the centers reported to the company's manager of development and the other 25 percent reported to the manager of information systems.

The main reasons for the growth of information centers are twofold:

a. Increasing widespread use of personal computers
b. End users wanting more access to information historically under the domain of data processing departments

The growth rate of the centers was 15 percent in 1983 and in 1984.

Spurred by user needs, Digital Equipment Corporation (DEC) announced in August 1984 at the Information Center Conference and Exposition that it would introduce the VAX Information Center concept, including VAX Toolbox, that would consist of a group of integrated software tools addressing a variety of end-user needs such as the ability to access databases, data analysis, and presentation of graphics, along with statistical modeling and query requirements. Knowing that it had to penetrate an IBM-dominated market, DEC made the VAX information center compatible with IBM mainframes and IBM personal computers. Thus, in a multivender environment it can run on both VAX and IBM hardware.

The FTP survey indicated the following as the greatest benefits coming out of information centers:

☐ Increased end-user computer literacy
☐ Improved end-user and MIS staff relationships
☐ More effective use of information systems resource
☐ Improved end-user productivity

The information center, however, has also faced a number of obstacles during its growth. Some of the problems are

☐ End users still have "computeritis," as one consultant called it, describing it as a fear of trying to learn something new, new ways of doing things, new equipment, new procedures and new responsibilities.

☐ Corporations' reluctance to invest capital without positive assurances of at least a dollar return for a dollar spent.

☐ Data processing professionals' fear of educating end users to the point where they, or their staffs, may be reduced when the end user becomes more sophisticated and is able to secure much of the information previously only provided through their domain.

☐ Difficulties in operating a user environment, developing and justifying a charge back, and identifying economics realized.

☐ Software products that were first available were insufficient, and newer products are unfamiliar.

The key to success of an information center is to identify the right manager who must be enthusiastic, persistent, optimistic, energetic, and innovative, as well as comprehensive and appreciative of the line of business and motivated for providing service to users. Cooperation between management and the MIS staff is essential for this person to succeed. The major factor influencing the future of information centers is the hidden data processing backlog in a company. The center manager must point out that the center is designed to reduce this backlog. Three factors will keep the center alive:

☐ Hidden backlog will continually demand attention.

☐ Some cases of success, where the right people are combined with the right software, will confirm the potential of information centers.

☐ Use of microcomputers will grow.

Eventually, the following scenario will emerge: End users will settle on the use of a microcomputer for many information center needs, and the information center specialist will shift focus to the delivery of data out of the corporate data resource to an information resource, which is fragmented over many microcomputers. The mainframe emphasis will be on a high-volume, high-performance transaction system having many integrated databases.

In an article published in *Datamation*, June 1, 1987, John N. Oglesky has proposed a "good info center shopping list" as follows:

1. A strategic plan that answers the two key questions
   ☐ What did my organization hope to accomplish when it started the information center?
   ☐ How can the information center best serve the organization?

2. A yearly plan of operation that determines the specific deliverables from the information center during the course of the year.

3. Appropriate hardware: if the center has to support the needs of small, independent work groups, then a set of microcomputers will be appropriate; if, on the other hand, large corporate databases need to be accessed by many different groups, then a mainframe will be necessary. Additional hardware related issues to be considered are the frequency of data transfer, reporting needs, and communication requirements.

4. Appropriate software: typical categories of software include fourth generation language, spreadsheets, DBMS, word processing, graphics, optimizing and nonoptimizing modeling, communications, and application generators.

5. Selection and training of staff of the center: the main difficulty here is that there are not enough people with the proper background and skills who can be attracted at the available salary.

6. End user training: this is vital because satisfied end users provide the greatest support of the information center.

7. Good evaluation and feedback program: the feedback mechanism is needed in order to find out how well or how poorly the center is functioning and in what ways it can be improved. The evaluation program will determine if the expectations of the end users are being met and if the objectives of the organization are being achieved.

15.3 INTERACTION AMONG MANAGEMENT, MIS STAFF, AND COMPUTER OPERATIONS STAFF

The key to the success of the information center concept lies in an effective and meaningful interaction among three separate groups of people in a company, namely, the management, the MIS staff, and the operations staff. In order to promote such communication, many researchers have proposed that two separate types of positions be created, one called the information specialist and the other the operations analyst.

The information specialist in a company should act as liaison librarian, interpreter, extractor, and evaluator, collecting mounds of material from a company's mini, micro, and mainframe computers and putting them into a format useful for executives. This person need not come from the data processing department but should have a clear understanding of the management's information requirements and a knowledge of the art of retrieving business information from technology as well as from personal contacts. It is quite possible that this person's role will clash with that of the MIS staff. However, the conflict can be avoided if this position is implemented properly and if the person does not play the role of a "mini-DP manager." The top management can alleviate this potential problem by clarifying what is the information specialist's turf and what is the MIS manager's turf.

An operations analyst, on the other hand, should play a liaison role between the MIS staff and the computer operations department. Normally, the operations people are somewhat looked down upon by the MIS staff who think of the former as maintenance people. But in reality the MIS staff are users of the computer resources that are managed by the operations staff who are again users of systems designed by the MIS staff. The operations analyst can keep the communication flowing in this cyclic environment. Edward Roeske has recommended in his recently published book (1983), *The Data Factory: Data Center Operations and Systems Development*, that such a position be created in every company in order to resolve problems and link the operations department with other data processing groups in the company.

Ideally, an operations analyst should work closely with the MIS staff and be involved in the total life cycle of a system. He or she is brought in at the system analysis stage to get a feel for what the system will do. During the detailed design stage the analyst scrutinizes the system file interfacing, as well as system scheduling and computer resource estimates. He or she then follows through with the implementation, unit testing, and system testing stages. This leads to a smooth project development.

When a company decides to install an information center, proper MIS planning is needed. Revalidation of plan from time to time by top management is essential. The information specialist and the operations analyst should work with the top management, MIS staff, and computer operations staff in order to allow the information center to function smoothly.

15.4 CONSULTANT HELP VERSUS IN-HOUSE EXPERTISE

Using consultants to identify potential problems is a common practice. When a company decides to hire consultants in search of rapid cost-effective solutions, it should determine beforehand that in-house expertise in that area is not available, because consultants are expensive and their liability or involvement after the solution is delivered is usually limited. This issue can be approached from three different points of view:

a. Staff perspective
b. Management's point of view
c. Consultant's point of view

From the staff perspective, the most common reaction is negative. The introduction of an "outsider" is often perceived as a lack of management's confidence in the staff's advisory capability. Further, the staff generally will have less commitment to the system's successful implementation than if it were of their own design. In addition, any area in which the consultant is duplicating previous or current staff activity will generally lead to the staff attempting to influence the consultant's results toward confirmation of the staff findings.

Not all staff reactions to the introduction of outsiders are negative. If the staff workload is significantly reduced by the contribution of the consultant or if the consultant's activities directly complement current or past staff activity, the reaction tends to be more positive. A positive reaction also appears in situations in which the consultant has sufficient credibility in the eyes of the staff to serve as an objective sounding board for new ideas or to provide an opportunity for education.

From the management point of view, the main emphasis is on getting the job done in the best currently acceptable manner. Three factors are involved here:

☐ Speed, which goes in favor of the consultant since he or she has access to information and resources that are not available to the company management. Also, the skills needed to do this job may not be possessed by the in-house staff.

☐ Expenses, which go in favor of the in-house staff members since they are already on the payroll and have to be paid regardless of the way the job is handled. Of course, if the in-house expertise is not available for the job due to other assignments, this factor leans toward the consultant.

☐ Clarity of solution, which is somewhat ambiguous. It favors the in-house staff if it has past experience in the area and has shown good performance. Otherwise, it helps the consultant who has established ability to handle the job.

Finally, from the consultant's point of view, money is normally the prime factor, which may be followed by the technical challenge offered by the job. In order to succeed in his or her role as a problem solver, the consultant must be technically competent and have the full support of top management, more so if he or she is perceived as a competitor by the in-house staff. Once introduced to the problem, the consultant should play an objective role remain detached from any alignment or polarization that may exist among the staff. Once the solution is reached, the consultant should submit the final recommendations as a report and exit the scene leaving all parties with a positive attitude and a feeling of having benefited from the experience.

Managers in the position to make the staff versus consultant decision have many issues s to consider, a primary one being solving the immediate problem. One way to ease the tension created by securing the services of an outside consultant is to permit the staff a voice in deciding who the consultant should be and to what degree the consultant will be involved.

Maintaining current knowledge and an air of objectivity appears to be the key to successful consulting experiences. Consultants should not hesitate to share the benefits of their experiences with the staff and should be willing to include the staff in their problem-solving approach.

In all fairness it should be pointed out that consultants are not as expensive as their hourly rates might suggest. For example, a consultant paid $50 an hour effectively earns $44,000 instead of $104,000 that is calculated by multiplying $50 by 2080 annual work hours. The reason is that consultants earn no salary on sick days, vacation days, or between projects. They also pay their own insurance, business expenses, and seminar fees. In-house staff, on the other hand, get paid between projects and are entitled to fringe benefits.

15.5   THIRD-PARTY LEASING OF EQUIPMENT

At the time of procurement of equipment during the system implementation phase the user organization faces three options: buy, rent, or lease. A detailed study is needed involving the lease versus purchase analysis, because rental of equipment is not so prevalent now. The user must select a residual value for the equipment. For example, if the equipment is worth 50 percent of its original value after 3 years, it is worth buying it. If, however, it is worth only 10 percent of its original value after 3 years, it is worth leasing it. However, if a company has already established a corporate policy (based mostly on accounting procedures) to buy or to lease, then the lease-versus-buy analysis is never carried out.

Third-party leasing often provides more attractive and less expensive options than the direct vendor can provide. Such options include the arrangement of financing and the upgrade of equipment. Users feel that third-party leasing saves them money since the third party assumes the residual risk of depreciation of the equipment value.

Third-party lessors are in the business of remarketing used equipment and leasing back new equipment. Hence they tend to be more flexible toward options for upgrading and assisting users in removing or terminating a system before the lease expires. When a vendor introduces a new model of equipment, it sells its old model, which is then purchased by the third-party lessors.

Buying always "locks" buyers into that equipment and so they have no chance of updating it when the next-generation equipment arrives on the market. The accelerated pull of technology often means that a company should lease rather than purchase.

At present about 90–95 percent of the computer leasing market involves IBM equipment. Researchers believe that this trend will continue for years. The reason for this IBM dominance is that IBM equipment is more frequently used than non-IBM systems. Also, third-party leasing involves the after-use market. The latter hardly exists for non-IBM equipment. A used computer dealer or broker may hesitate to take any equity position in non-IBM equipment because the chance of being undercut can be strong. This results in less liquidity in the non-IBM market, leading to an active secondary market in IBM equipment.

The relationship between IBM and third-party lessors began in the 1960s, when the lessors entered the computer market. At that time IBM was renting most of the equipment it produced. During the 1970s lessors' businesses grew and IBM continued its practice of renting computers on a month-to-month basis. A major change in the relationship between lessors and IBM occurred in 1981, when IBM established IBM Credit Corporation (ICC).

IBM Credit Corporation leased equipment, rather than rented it (as did IBM itself), and thus became a direct competitor with third-party lessors. Although IBM continued to rent equipment, it increasingly moved away from this business.

In 1981, IBM rented two-thirds of its equipment and sold one-third. Today, those figures are reversed and by the early 1990s IBM will probably sell all but 1 or 2 percent of its equipment. Meanwhile, ICC has grown to become one of the leasing industry's top lessors, with revenues exceeding $100 million. Despite this competition, business is booming for third-party lessors. Industry revenues were up 28 percent in 1983 and are expected to grow substantially in 1984.

Lessors deal primarily with peripheral equipment, because there is less risk, less fear of obsolescence. Customers do not evaluate peripherals as carefully as computers.

It should be noted, however, that some market researchers believe that non-IBM leasing will be growing in the future. For example, Digital Equipment, Data General, and Hewlett-Packard systems currently play a dominant role in the non-IBM market. Leasing non-IBM equipment on a short-term basis, say upto 3 months, is becoming popular. Often customers who have ordered DEC or DG equipment lease them from third-party lessors for a short period of time until their purchased equipment arrives.

15.6 NEW METHODOLOGY FOR STRUCTURED DEVELOPMENT

In an article, "What Ever Happened to Structured Analysis?" published in *Datamation* ([6]) Edward Yourdon wrote the following:

> Many organizations made a valiant effort to use structured analysis but eventually abandoned it altogether. Others spent a long time incorporating structured analysis into standards manuals only to find, years later, that no one followed it. Still others who used structured analysis happily in the late 1970s and early 1980s became convinced later on that tools like fourth generation languages and prototyping packages were better.
>
> What happened to the revolutionary fervor with which structured analysis was accepted a relatively few short years ago?

Yourdon has offered three main reasons for the current "lack of interest" in structured analysis techniques:

1. A real-world project requires a large number of data flow diagrams for documentation of the logical system. Usually each diagram is revised several times before being finalized. This imposes a substantial amount of manual labor on the system developers.

2. Classical structured analysis is unable to handle complex real-time systems. Little or no guidance is provided to system developers who build systems involving networks of mainframes, minis, micros, and smart on-line terminals.

3. Many prospective users of the structured techniques have been lured away by the promises of fourth-generation languages (4GL), prototyping tools, and application generators.

This does not mean, however, that the structured techniques are obsolete. Since their inception in the late 1970s, they have been evolving and people's attitudes toward them have been changing. More and more end users, top management, and technical personnel are demanding top quality, reliable, and maintainable systems for which structured methodology is necessary. On the other hand, 4GL and prototyping tools have been somewhat oversold. In order to apply the structured methodology to current day real-time systems, it is necessary to change its orientation.

There are three main dimensions of complexity for a real-time system: functions, data, and time-dependent behavior. Traditional structured techniques address the functional complexity by means of the data flow diagrams. Recent trends suggest that data complexity should be handled by means of entity-relationship diagrams (see [1]). Each data store in a data flow diagram should correspond to an entity in an entity-relationship diagram and vice versa. Finally, the time-dependent behavior should be modeled via state-transition diagrams. During the development of the logical system, it should be checked if the data flow diagrams and the state-transition diagrams are consistent with each other. In general, we need cross-checking rules to ensure that the three separate complexity entities do not conflict with one another. Yourdon has commented as follows ([6], p. 136):

> Increasingly, we are finding that big systems have complexities in all three dimensions. Banking and insurance systems, for example, not only have increasingly complex data and functions (which must be modeled accurately during the systems analysis phase), but also have thousands of on-line terminals with response-time requirements that are becoming more and more stringent. Similarly, today's real-time systems often contain hundreds of thousands, if not millions, of lines of code to carry out increasingly complex functions.
>
> The tools of structured analysis have evolved to model real-time systems. Early textbooks' simple data structure diagrams which detail data and their component

attributes, have been replaced by more appropriate entity-relationship diagrams, which depict relationships among high-order groupings of data, i.e., entities. The modeling of time-dependent behavior is now carried out with state-transition diagrams, which depict the various states of a system and conditions that lead to a change from one state to another.

We expect that the Computer Aided Software Engineering (CASE) technology will help revolutionize the software industry. In 1986 there were about 6000 CASE workstations installed in the United States. This figure translates to nearly 1 percent of programmers and analysts. By 1990, nearly 10 percent of programmers and analysts should be equipped with CASE and by 1995 that figure should grow to 90 percent. If this trend continues, more vigor will be injected into the structured technique, leading to its future growth. The present generation of top and middle management realizes that it cannot ignore the issues of building capable information systems because its own companies' strategies and abilities to compete depend on such systems. Therefore, it is absolutely essential that reliable and maintainable models of system requirements be built. The revised form of structured methodology can accomplish this goal with the help of the CASE technology or other similar software that help automate the system analyst's job.

15.7 ON-LINE TRANSACTION PROCESSING*

On Line Transaction Processing (OLTP) represents an application in which a common database is available to interactive terminal users for both inquires and updates. The database contains all the data used for the application. A transaction represents an activity that causes an update of the database (e.g., receipt of a shipment, change of a customer name) or involves data retrieval from the database in consequence of an inquiry.

The airline reservation system introduced in the mid-1960s was the first application using OLTP. Additional applications were started during the mid-1970s, some of which are

☐ Hotel/motel reservations
☐ Car rentals
☐ Large-scale banking applications such as demand/deposit teller support, automatic teller machines, and electronic fund transfer
☐ Credit authorization systems
☐ Trading and brokerage systems

*This section is based on the article, "Exploring the OLTP Realm," by Omri Serlin, published in *Datamation*, August 1985, see [5].

IBM hardware is used in the majority of OLTP systems. In 1982, 1983, and 1984 the percentages of applications using IBM hardware were 57, 59, and 57, respectively. The percentage is expected to increase in the future. The Customer Information Control System (CICS) package is used in nearly 70 percent of the IBM installations to implement an OLTP.

Major future applications with a varying degree of OLTP include

1. **Airline reservations.** Airline reservations remain the predominant type of OLTP systems. In these systems the common database, representing the inventory of available seats on various flights, is made accessible to travel agents using interactive terminals. The agents can make various inquiries against the database (such as which flights operate between two cities on a given data on or about a given hour and what class seats are available). Then the agent can make an actual booking, which involves on-line database updates reducing the number of available seats on a given flight by the number booked and creating a related passenger name record (PNR) containing information such as the passenger's name and phone number. A booking cancellation also results in a database update.

2. **Banking and financial applications.** The most visible banking and financial application provides support to tellers' activities (e.g., automatic demand/deposit and electronic funds transfer). For example, in 1977 the Bank of America installed a multiple minicomputer system to help 10,000 tellers in 1200 branches administer 10 million customer accounts. In each of the bank's two processing centers (Los Angeles and San Francisco), a network of 32 minicomputers acts as a front end to large IBM mainframes, where account information is held. Every morning the account status is loaded into the minisystem; during the day, most teller transactions involving debits and credits to the accounts are merely captured on line. At night, the transactions are processed in batch fashion by the main computers to update the account information. There is a need to maintain absolute accuracy (in account balances, for example), despite concurrent update activities and, possibly, hardware failures.

3. **Manufacturing.** Increasing attention is being focused on shop floor control and other systems involved in the minute-by-minute operation of manufacturing plants. The concept of a paperless factory is rapidly gaining acceptance. In such factories, the paperwork accompanying each product of batch is replaced by a combination of bar-code stickers and information entered through terminals located in the various work areas. The systems supporting paperless factories are prime examples of the application of OLTP technique in the manufacturing environment.

Because of increasing concerns regarding industrial productivity, such systems have been funded more aggressively during the early 1980s and are likely to continue to be the focus of substantial interest. These systems save time that was once lost to form filling; provide up-to-the-minute reports on the status of plant, inventories, and finished goods; and can be used to pinpoint sources of production problems earlier than before.

An OLTP system should be capable of handling the following issues:

a. Fault tolerance and availability.

Since OLTP systems frequently govern the operation of a company, a disruption in its operation can have serious consequences. An airline with a malfunctioning reservation system cannot operate. A bank with an inoperative automated teller machine creates irate customers.

Some of the key ingredients of today's (1987) fault tolerant systems were first developed and tested in the airline reservations systems. Among these are disk mirroring, or shadowing, a scheme by which a system undertakes to maintain a mirror image of critical disk data by automatically replicating every write to a second, physically separate drive. This assures that the critical portions of the database are physically accessible even after a disk drive failure.

Another common problem associated with fault-tolerant systems is the concurrent access and update of the common database used by the system. The most common solution for this type of problem is to serialize the concurrent accesses by means of file and record locks. Thus, if an agent's transaction intends to modify the seat inventory, a lock is placed on the relevant flight record to prevent anyone else with similar update intent from reading the seat status until the first agent is through.

A third problem related to fault tolerance is the exactness of results. For example, a bank must tell its customers their exact account balances, not approximate ones. The concept of atomic transactions originated to handle this problem. Atomicity is a property of the system under which users may define a sequence of related database actions (by, for example, bracketing the sequence with BEGIN TRANSACTION and COMMIT commands); the system undertakes to guarantee either that all such actions will be executed or that none will.

There are various ways to achieve this guarantee. Perhaps the most popular is the write-ahead log technique in which the before and after images of the affected records are first written to a mirrored disk log file. If the system is successful in implementing these changes in the actual database, a success record is added to the log file.

If the processor or the system crashes in the middle of such actions, the system can examine the log file when it is restarted; those transactions not successfully implemented (detected by the absence of the success record in log file) can be undone by using the before and after images. As long as one copy of the log file and the database remain intact, any number of system failures can be compensated for by repeating this transaction-backout process, which removes the effects of incomplete transactions. Tandem Computers of Cupertino, California, and Stratus Computer of Marlboro, Massachusetts, are two prime suppliers of fault-tolerant systems for commercial OLTP applications.

b. Performance considerations. An appropriate measure of OLTP performance is transactions per second, or tps. The cost per tps for typical multiprocessor systems such as Tandem or Stratus is about $50,000 to $100,000 per tps. Such systems can deliver about 1 to 5 tps per CPU under this benchmark. Large IBM mainframes can do perhaps 10 times better in terms of tps/CPU, but they are not necessarily more cost effective in terms of cost per tps. Unfortunately, not enough comparative data are available. Large OLTP customers such as airlines, financial institutions, and brokerage firms foresee the need for systems offering 1000 or more tps. Since a loss of performance occurs due to multiprocessing overhead, which results in a less than linear performance increase as the number of processors is increased in a multiprocessing environment, it is not yet known whether architectures that rely on multiple low-powered processors can effectively provide a high level of tps.

15.8 SYSTEM DEVELOPMENT FOR EXPERT SYSTEMS

In this section we discuss how the five-phase system development process can be used to design and implement an expert system.

15.8.1 Concept of Expert System

An expert system has been defined as follows by Forsyth ([2], p. 10):

> An *expert system* is regarded as the embodiment within a computer of knowledge-based component, from an expert skill, in such a form that the system can offer intelligent advice or make an intelligent decision about a processing function. A desirable additional characteristic, which many would consider fundamental, is the capability of the system, on demand, to justify its own line of reasoning in a manner directly intelligible to the enquirer. The style adopted to attain these characteristics is rule-based programming.

An expert system has four components:

a. Knowledge base
b. Inference engine
c. Knowledge acquisition module
d. Explanatory interface

The knowledge base contains a set of rules, usually given in a condition-action format using an IF-THEN type construct. It is a dynamic entity that is kept up-to-date by means of the knowledge acquisition module. The latter captures new knowledge in the area of the expert system and incorporates it in the knowledge base. Care is taken to ensure that the new rules do not bring any inconsistency or contradiction into the knowledge base. The explanatory interface is the user-friendly front end for the expert system and is often called a "human window" to the system. It extracts information from the user through prompts and passes it on to the inference engine. The latter may be compared to the human brain in that it executes the reasoning process in a forward or backward chaining form. The inference engine interacts with the knowledge base by matching the information provided by the user with the production rules contained in the knowledge base. During this interaction the user may be prompted for additional information. Before supplying additional evidence, the user may ask *why* the said evidence is needed. Similarly, at the end of the session with the expert system, the user may ask the system *how* it arrived at its conclusion. Figure 15-1 shows the user interaction with the expert system. Note that except for the explanatory interface, everything else is user transparent. The "conversation" with the user takes place in plain English. Thus, the explanatory interface includes a natural language understanding scheme. Normally, the keyword matching technique is used to interpret the user's plain English input.

15.8.2   Five-Phase Development Process

The five phases of structured system development process (see Chapter 2) are:

a. Problem definition and feasibility
b. System analysis
c. Preliminary system design
d. Detailed system design
e. System implementation, evaluation, and maintenance

The five phases of expert system development can be identified as follows ([3], pp. 34–35):

a. Identification. Some tasks in a specific domain that are normally done by a human expert are selected as target candidates for an expert system. These tasks must be feasible in terms of available hardware and software.
b. Knowledge acquisition. The expert's knowledge has to be extracted and represented in some form so that a logical system evolves. The expert then reviews this system to verify that it indeed is a reasonable

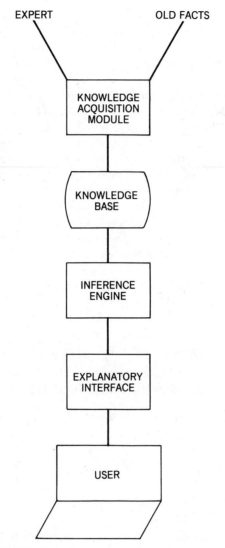

FIGURE 15-1 Components of one expert system.

conceptual representation of his or her domain. At this stage, the logical system should include the following issues

☐ What are the inputs or problems?
☐ What are the outputs or solutions?
☐ Which types of inputs cause difficulties for the expert?
☐ How are the problems characterized?
☐ How are the solutions characterized?

☐ What sort of knowledge is used?
☐ How are problems or methods broken down into smaller units?

A later, more detailed, breakdown would answer detailed questions such as the following:

☐ What data are input; in what order and form?
☐ What are the relationships between data items?
☐ How important and how accurate are the data items?
☐ Which data might be missing?
☐ What assumptions does the expert make?
☐ What constraints does the expert have?
☐ What sort of inference does the expert make?
☐ How does the expert form concepts and hypotheses?
☐ How do these relate to each other?
☐ How does the expert move from one state of belief to another?
☐ Which evidence suggests particular goals or concepts?
☐ What are the causal relationships?
☐ Are there any logical constraints on the system?
☐ Which problems are easy, common, hard, interesting, etc.?

c. **Design.** Once the expert is satisfied with the logical system, the problem of system design is addressed. The outcome of this phase is the physical system. This involves choosing appropriate structures to represent the knowledge base and inference engine as they will appear within the expert system.

d. **Development and testing.** During this phase the designed system is implemented. All aspects of the system are tested. The process of testing can be difficult due to the exploratory nature of the expert system. Often it is found that appropriate hardware is not available for large expert systems that require a high speed of rule firing. The relationship between complexity of implementation and the number of rules in the knowledge base is not linear. Thus, an expert system using 500 rules is more than 10 times as complex to implement than an expert system with 50 rules.

e. **Use and evaluation.** The expert system must be used cautiously at first. It should be under continuous review and evaluation for some time before it is used with confidence.

15.8.3 Analogy between System Analysis and Knowledge Acquisition

A system analyst performs the following sequence of activities:

a. Interviews the users and reviews current operations manual as well as earlier studies to understand the present system

b. Formulates a set of functional requirements of the new system

c. Validates the findings with the user

The knowledge engineer plays a role similar to that of the system analyst during the knowledge acquisition process. The knowledge engineer's activities can be summarized as follows:

a. Interviews the expert to gain an understanding of the expert's domain of knowledge

b. Poses specific questions to the expert to elicit knowledge

c. Codifies the elicited knowledge in a format ready to be included in the knowledge base

d. Verifies from the expert that the coded rules are a valid representation of the expert's knowledge

It is clear from above that a system analyst and a knowledge engineer play similar roles, as do the system user and the expert. The system user is knowledgeable in his or her operations but not in system design or programming. The system analyst plays the liaison between the users' operations and the system design. Likewise, the expert is knowledgeable in his or her domain but not in the representation of the knowledge as coded rules. The knowledge engineer provides the link between the expert's knowledge and the coded rules. Figure 15-2 gives a graphic representation of this situation.

Both the system analyst and the knowledge engineer must possess similar skills to succeed in their respective jobs. Some of the needed skills are

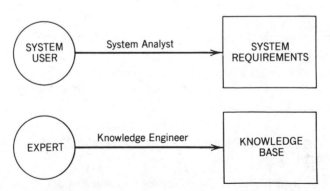

FIGURE 15-2   System analyst versus knowledge engineer.

☐ Good communication ability

☐ Aggressive desire to learn

☐ Tact and diplomacy

☐ Empathy and patience

☐ Persistence

☐ Logical bent of mind

☐ Innovativeness

☐ Self-confidence

☐ Domain knowledge

☐ Programming knowledge

Fact finding techniques discussed in Chapter 3 apply equally well to system analysis and knowledge acquisition.

15.8.4 Analogy between System Design and Knowledge Representation

System analysis and system design are distinct activities. The latter takes into consideration the available hardware and software in designing a physical system. These constraints should have been spelled out by the feasibility study done earlier. In the same way, knowledge representation is different from knowledge acquisition. The knowledge engineer has to anticipate how a particular inference method or knowledge base can be physically represented for later implementation. Consequently, after the knowledge acquisition phase is complete, the knowledge engineer must formulate the findings in proper format.

Rules in a knowledge base can be coded in a variety of ways, such as production system, semantic network, or frame ([4], Chapters 5 and 7). The knowledge engineer decides upon the particular method and then converts the information gathered earlier into the appropriate form. Several alternatives are available at this stage:

a. **Tabular representation of knowledge.** The knowledge engineer formulates rules by using facts and axioms. The expert system matches the input data with appropriate rules to give a resulting decision. The inference engine contains algorithms describing how the input data are used to obtain output decisions.

b. **Knowledge representation by example.** The knowledge engineer uses existing documentation such as graphs and charts to formulate rules and correlate them with these examples. Input data are matched with the examples to arrive at decisions.

c. **Inference network.** The network consists of a set of arcs and nodes. A node represents facts, and arcs leading to a node represent rules. For example, Figure 15-3 represents the rule

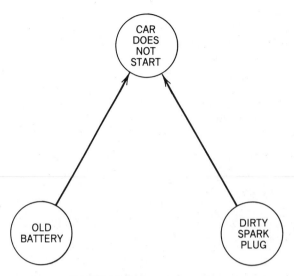

FIGURE 15-3 Interfase network.

```
IF OLD BATTERY AND DIRTY SPARK PLUGS,
THEN CAR MAY NOT START.
```

The network can become very complicated with a large number of rules.

d. Decision tree structure. A decision tree is similar to a flowchart but consists of nodes and arcs. Terminal nodes are those at the bottom; the others are intermediate nodes. In this case, we branch down the tree along a path depending on the value of an attribute described in an intermediate node until a terminal node (the decision) is reached. The rule is formulated by starting from a top node and following a path until we reach a terminal node.

The expert and the knowledge engineer should agree on a mode of representation of knowledge that is mutually acceptable and useful. The choice may change as the project gets underway, and a fresh start may be needed. However, the sooner the results can be represented and documented, the better. This helps the expert maintain interest and enthusiasm and gives him or her something on which to comment.

15.9 SUMMARY

This chapter discussed the following seven heterogenous topics that are relevant to the total system development process:

1. Information center concepts
2. Interaction among management, MIS staff, and operations staff
3. Consultant help versus in-house expertise
4. Third-party leasing of equipment
5. New methodology for structured development
6. On-line transaction processing
7. System analysis for expert systems

An information center is regarded as a part of the MIS department. Its consultants work with users and provide them with necessary services such as reports. This concept existed during the mid 1970s but has been heavily promoted by IBM only since the early 1980s. Its popularity is increasing. As of 1987, Digital Equipment Corporation has introduced the VAX Information Center in order to capture the market. However, due to a lack of communication among the different departments within a company, sometimes the information center faces obstacles. The key to success for this concept is to have a good manager in charge of the center and to have a proper support and cooperation from management.

Related to the success of an information center is the need for good communication among the management, the MIS staff, and the operations staff. To facilitate cooperation, two types of positions have been proposed: information specialist and the operations analyst. The information specialist acts as a liaison between the technical MIS department and the management. He or she should have a clear understanding of the management's information requirement and a knowledge of the art of retrieving business information from technology as well as from personal contacts. An operations analyst plays a liaison role between the MIS staff and the computer operations department. He or she should work closely with the MIS staff and be involved in the total life cycle of a system.

From time to time a company may face the decision between hiring a consultant versus using in-house expertise to solve a problem. This issue can be approached from a staff perspective, management's point of view, or a consultant's point of view. A consultant is often hired because of special expertise and the speed with which he or she can solve the problem. However, a consultant may be treated as an adversary by the in-house staff, thus, management support is crucial to enable a consultant to succeed. The expenses associated with a consultant's services are often regarded as a negative factor.

At the time of procurement of equipment during system implementation, third-party leasing of equipment, especially the peripherals, is a viable alternative. About 90 to 95 percent of third-party leasing involves IBM equipment. Third-party lessors are in the business of remarketing used equipment and leasing back new equipment. Hence they are more flexible in options for upgrading and assisting users in removing or terminating a system before the lease expires. Some market researchers think that the non-IBM lease market will grow in the future.

The structured method of system development t is found to be somewhat inadequate to handle today's on-line/real-time systems. Consequently, Yourdon has proposed a new methodology that can address three main dimensions of complexity of a real-time system: functions, data, and time-dependent behavior. Data flow diagrams of structured analysis handle the functional complexity. Entity relationship diagrams of Chen deal with the data relational complexity. State-transition diagrams address the time-dependent behavior of real-time systems. The Computer Aided Software Engineering (CASE) technology uses this approach and is expected to revolutionize the software industry.

On Line Transaction Processing (OLTP) represents an application in which a common database is available to interactive terminal users for both inquiries and updates. The airline reservation system was the first application using OLTP. In 1986, a variety of applications used OLTP. Examples include car rentals, banking, credit authorization, trading and brokerage. Systems using OLTP should have a high level of fault tolerance and be available almost around the clock. Its efficiency is measured in terms of transactions per second. Large-scale OLTP systems may need computers performing at a rate of 1000 tps or more.

The development of an expert system uses the five phases of the structured methodology, although in a different setting. Knowledge acquisition is analogous to system analysis. A knowledge engineer interacts with an expert and gathers the necessary information in the expert domain. Standard interviewing techniques are followed for that purpose. Knowledge representation is analogous to the system design. In this phase the knowledge engineer codes the knowledge in appropriate rule format for future implementation. The development and testing of expert systems are more difficult than the implementation of business applications because of the exploratory nature of the expert systems.

15.10 KEY WORDS

The following key words are used in this chapter:

Computer Aided Software Engineering (CASE)	expert
	expert system
consultant	explanatory interface
data flow diagram	fault tolerance
decision tree structure	human window
end user	inference engine
entity-relationship diagram	inference network

information center

information specialist

in-house expertise

knowledge acquisition

knowledge base

knowledge engineer

knowledge engineering

knowledge representation

logical system

node

On Line Transaction Processing (OLTP)

operations analyst

performance

physical system

real-time system

state-transition diagram

structured analysis

system life cycle

third-party leasing

third-party lessor

transaction

transactions per second (tps)

VAX information center

VAX toolbox

REFERENCES

1. P. P. Chen, *Entity Relationship Model*, ACM Transactions on DBMS, vol. 1, no. 1, March 1976, pp. 9–36.
2. R. Forsyth (ed.), *Expert Systems*, Chapman and Hall Computing, New York, 1984.
3. A. Hart, *Knowledge Acquisition for Expert Systems*, McGraw Hill, New York, 1986.
4. E. Rich, *Artificial Intelligence*, McGraw Hill, New York, 1983.
5. O. Serlin, "Exploring the OLTP Realm," *Datamation*, August 1985.
6. E. Yourdon, "What Ever Happened to Structured Analysis?" *Datamation*, June 1, 1986, p. 133–138.

INDEX